Thurston's Work on Surfaces

Thurston's Work on Surfaces

Albert Fathi, François Laudenbach, and Valentin Poénaru

Translated by Djun M. Kim and Dan Margalit

Mathematical Notes 48

PRINCETON UNIVERSITY PRESS
PRINCETON AND OXFORD

First published in France under the title *Travaux de Thurston sur les surfaces*, 1979

Published with the permission of the Société Mathématique de France

English edition copyright © 2012 by Princeton University Press

Published by Princeton University Press, 41 William Street,
Princeton, New Jersey 08540

In the United Kingdom: Princeton University Press, 6 Oxford Street,
Woodstock, Oxfordshire OX20 1TW

press.princeton.edu

Library of Congress Cataloging-in-Publication Data

Travaux de Thurston sur les surfaces. English.
   Mathematical notes/[Edited by] Albert Fathi, François Laudenbach, and Valentin
Poénaru; translated by Djun Kim and Dan Margalit.
      p. cm. — (Mathematical notes ; 48)
   Includes bibliographical references and index.
   ISBN 978-0-691-14735-2 (pbk.)
      1. Homeomorphisms. 2. Surfaces. 3. Dynamics. I. Fathi, Albert. II. Laudenbach,
François. III. Poénaru, Valentin. IV. Title.
QA614.T7313 2011
514—dc23

                              2011029588

British Library Cataloging-in-Publication Data is available

The publisher would like to acknowledge Djun M. Kim and Dan Margalit
for providing the camera-ready copy from which this book was printed

This book has been composed in Computer Modern Roman in LaTeX

Printed on acid-free paper ∞

Printed in the United States of America

10 9 8 7 6 5 4 3 2

Contents

Preface

FOREWORD TO THE FIRST EDITION

This book contains an exposition of results of William Thurston on the theory of surfaces: measured foliations, the natural compactification of Teichmüller space, and the classification of surface diffeomorphisms. Our scope is essentially that outlined in the research announcement of Thurston and in the notes of his Princeton course, as written up by M. Handel and W. Floyd.

Part of this work, most notably the classification of curves and of measured foliations, is an elaboration of lectures made in the Seminaire d'Orsay in 1976–1977. But we were not able to write the proofs for the remaining portions of the theory until much later. In the Spring of 1978, at Plans-sur-Bex, Thurston explained to us how to see the projectivization of the space of measured foliations as the boundary of Teichmüller space.

The first exposé enumerates the principal results, the proofs of which follow in Exposés 2 through 13. The last two exposés present work somewhat marginal to the theme of the classification of surface diffeomorphisms. Exposé 14, orally presented by D. Fried and D. Sullivan, discusses nonsingular closed 1-forms on three-dimensional manifolds, following Thurston; in particular it treats fiber bundles over S^1 for which the monodromy diffeomorphism is pseudo-Anosov. Exposé 15, presented orally by A. Marin, gives a finite presentation of the mapping class group, following Hatcher and Thurston.

The seminar consisted also of exposés of an analytical nature (holomorphic quadratic differentials, quasiconformal mappings) presented by W. Abikoff, L. Bers, and J. Hubbard. In the end, the two points of view were found to be more independent of each other than was initially believed. The analytic point of view is the subject of a separate text written by W. Abikoff; see [Abi80].

We thank all of the active participants in the seminar; all have contributed assistance in various sections: A. Douady, who, after the oral presentations, helped us to capture the content of the lectures; M. Shub, who discussed with us the ergodic point of view; D. Sullivan, who, besides giving much advice and encouragement, strove to make us understand how the image of a curve under iteration of a pseudo-Anosov diffeomorphism "approaches" a foliation of the surface (it took many more months to fully understand this "mixing").

Finally, we thank Mme. B. Barbichon (typing) and S. Berberi (illustrations) for the care that they took in preparing the manuscript.

A. Fathi – F. Laudenbach – V. Poénaru

FOREWORD TO THE SECOND EDITION

The research announcement "On the geometry and dynamics of diffeomorphisms of surfaces," by William Thurston ([Thu88]), has finally appeared in the Bulletin of the American Mathematical Society. One can find there a list of references to later work and to the first edition of this book. We also point out the book ([CB88]) by A. Casson and S. Bleiler, "Automorphisms of Surfaces after Nielsen and Thurston."

We limit ourselves to a few corrections that one can find assembled in the errata at the end of this volume.[1]

Orsay, May 27, 1991   FLP

---

[1]Translators' note: We have incorporated these corrections into the main text of the translated edition.

TRANSLATORS' NOTES

We are very happy to present this translation of the now-classic text *Travaux de Thurston sur les surfaces* ([FLP79]), commonly referred to as *FLP*. We have attempted to stay as faithful to the original as possible. At the same time, we have made many small modifications in order to elucidate the structure of the theory, modernize the language, and clarify the details of some definitions and proofs.

In the three decades since its original publication, *FLP* was the only source for many of the details involved in the measured foliations point of view of Thurston's work. However, several other books and papers have appeared that elucidate other aspects of Thurston's theory, for instance, those of geodesic laminations, train tracks, $\mathbb{R}$-trees, geodesic currents, and quadratic differentials. The following works give an overview of these perspectives.

We hope that this translation inspires the reader to learn about some of these more modern developments.

Finally, corrections and additional material may be found at our website http://flpbook.org/.

"An Extremal Problem for Quasiconformal Mappings and a Theorem by Thurston," Lipman Bers, *Acta Mathematica*, 1978. ([Ber78])

The Real Analytic Theory of Teichmüller Space, William Abikoff, Springer, 1980. ([Abi80])

"New Proofs of Some Results of Nielsen," Michael Handel and William P. Thurston, *Advances in Mathematics*, 1985. ([HT85])

"On the geometry and dynamics of diffeomorphisms of surfaces," William P. Thurston, *Bulletin of the American Mathematical Society*, 1988. ([Thu88])

Automorphisms of Surfaces after Nielsen and Thurston, Andrew J. Casson and Steven A. Bleiler, Cambridge, 1988. ([CB88])

"The Geometry of Teichmüller Space via Geodesic Currents," Francis Bonahon, *Inventiones Mathematicæ*, 1988. ([Bon88])

Subgroups of Teichmüller Modular Groups, Nikolai V. Ivanov, AMS, 1992. ([Iva92])

"Train-tracks for surface homeomorphisms," Mladen Bestvina and Michael Handel, *Topology*, 1995. ([BH95])

"$\mathbb{R}$-trees in topology, geometry, and group theory", Mladen Bestvina, in *Handbook of geometric topology*, North-Holland, 2002 ([Bes02])

Teichmüller Theory and Applications to Geometry, Topology, and Dynamics, Volume I: Teichmüller Theory, John H. Hubbard, Matrix Editions, 2006. ([Hub06])

A Primer on Mapping Class Groups, Benson Farb and Dan Margalit, Princeton University Press, 2011. ([FM11])

ACKNOWLEDGMENTS

We offer our sincerest thanks to Jayadev Athreya, Hyungryul Baik, Matt
Bainbridge, Jessica Banks, Jason Behrstock, Aaron Brown, Jeff Carlson, Indira
Chatterji, Matt Clay, Moon Duchin, Craig Hodgson, Keiko Kawamuro, Justin
Malestein, Ben McReynolds, Erika Meucci, Takuya Sakasai, Saul Schleimer,
Dylan Thurston, Liam Watson, Saadet Yurttas, and the referees for reading
early drafts and offering corrections and helpful suggestions. We are especially
grateful to Allen Hatcher, Chris Leininger, and Lee Mosher for several
in-depth conversations.

Djun would like to thank Dale Rolfsen, for providing support, encouragement,
friendship, and many fruitful discussions over the years; the Mathematics
Departments at Berkeley and McGill, for providing library and computer
facilities during visits in 1990 and 1996 respectively, and the Mathematics
Department at the University of British Columbia, for providing resources and
a lively and collegial working environment. He is grateful to Elisabeth Kim,
who read early drafts of some of the exposés and provided corrections and
helpful suggestions, and to Bill Casselman, Curt McMullen, Kevin Pilgrim,
Larry Roberts, and Jonathan Walden for encouragement and advice. Finally,
he thanks Pamela Richardson, for her support and encouragement in the final
years of the preparation of this book.

Dan would like to thank the University of Chicago, the University of Utah,
Tufts University, the Georgia Institute of Technology, and the National Science
Foundation for providing resources and pleasant working environments. He is
also grateful to Benson Farb, Mladen Bestvina, Joan Birman, and Kevin
Wortman for their invaluable mathematical and moral support. He would like
to thank his family for their endless encouragement, and most of all his wife,
Kathleen, for her constant love, support, and inspiration.

We are most grateful to Vickie Kearn and Mark Bellis, our editors, and Anna
Pierrehumbert and Stefani Wexler, Editorial Assistants/Associates at
Princeton University Press. Carole Schwager's meticulous attention to detail
and style have been invaluable.

Finally, we would like to thank William Thurston, as well as Albert Fathi,
François Laudenbach, Valentin Poénaru, and all others who contributed to the
beautiful mathematics contained in this volume.

Djun M. Kim   Dan Margalit
Vancouver, BC, Canada   Atlanta, GA
Djun.Kim@math.ubc.ca   Margalit@math.gatech.edu

ABSTRACT

This book is an exposition of Thurston's theory of surfaces: measured foliations, the compactification of Teichmüller space, and the classification of diffeomorphisms. The mathematical content is roughly the following.

For a surface M (let us say closed, orientable, of genus $g > 1$), one denotes by $\mathcal{S}$ the set of isotopy classes of simple closed curves in M. For $\alpha, \beta \in \mathcal{S}$, one denotes by $i(\alpha, \beta)$ the minimum number of *geometric* intersection points of α' with β', where α' (resp. β') is a simple curve in the class α (resp. β). This induces a map $i_* \colon \mathcal{S} \to \mathbb{R}_+^{\mathcal{S}}$ which turns out to be injective. In fact, if one projectivizes $\mathbb{R}_+^{\mathcal{S}} \setminus 0$, then i_* induces an injection $i_* \colon \mathcal{S} \to P(\mathbb{R}_+^{\mathcal{S}})$ which endows $\mathcal{S}$ with a nontrivial topology. Here $\mathbb{R}_+^{\mathcal{S}}$ is endowed with the weak topology (= product topology). Two curves $\alpha, \beta \in \mathcal{S}$ are "close" in $P(\mathbb{R}_+^{\mathcal{S}})$ if, up to a multiple, they are made up of more or less the same strands going in more or less the same direction; this is very different from saying that the curves are homotopic.

The limits of curves are naturally interpreted as projective classes of "measured foliations," that is, foliations that have an "invariant" transverse distance, and that have certain kinds of singularities (well-known in the theory of quadratic differentials, or in smectic liquid crystals). The space of measured foliations considered in $\mathbb{R}_+^{\mathcal{S}}$ (or in $P(\mathbb{R}_+^{\mathcal{S}})$) is denoted by $\mathcal{MF}$ (resp. $\mathcal{PMF}$). One shows that

$$\mathcal{MF} \simeq \mathbb{R}^{6g-6} \quad \text{and} \quad \mathcal{PMF} \simeq S^{6g-7}.$$

In $P(\mathbb{R}_+^{\mathcal{S}})$, the space $\mathcal{PMF}(M)$ and the Teichmüller space $\mathcal{T}(M)$ glue together into a $(6g - 6)$-dimensional ball:

$$\overline{\mathcal{T}}(M) = \mathcal{T}(M) \cup \mathcal{PMF}(M) = D^{6g-6}.$$

The group $\mathrm{Diff}(M)$ acts continuously on this compactification of $\mathcal{T}$ (this is therefore a "natural" compactification). Thus any $\phi \in \mathrm{Diff}(M)$ has a fixed point in $\overline{\mathcal{T}}(M)$ (Brouwer) and the analysis of this fixed point shows that (up to isotopy) each ϕ is either a hyperbolic isometry of M, is "Anosov-like" (the word is "pseudo-Anosov"), or else is "reducible." Pseudo-Anosov diffeomorphisms minimize the topological entropy in their isotopy class. Also, two pseudo-Anosov diffeomorphisms that are isotopic are actually conjugate.

Every diffeomorphism $\phi \colon M \to M$ has a (finite) spectrum defined in terms of the length of $\phi^n(\alpha')$ raised to the power $1/n$. A pseudo-Anosov diffeomorphism is characterized by the fact that the spectrum is a single value $\lambda > 1$.

There is a good method for producing many pseudo-Anosov diffeomorphisms out of combinations of Dehn twists, which is explained in Exposé 13.

The last two exposés are of a somewhat different character: Exposé 14 is about closed nonsingular 1-forms on 3-manifolds, and Exposé 15 is about the Hatcher–Thurston theorem on finite presentability of the mapping class group, $\pi_0(\mathrm{Diff}(M))$.

Thurston's Work on Surfaces

Exposé One

An Overview of Thurston's Theorems on Surfaces

by Valentin Poénaru

1.1 INTRODUCTION

Thurston's theory ([Thu88], see also [Thu80], [Poé80]) is concerned with the following three problems:

> 1   Describing all simple closed curves on a surface up to isotopy.

> 2   Describing all diffeomorphisms of a surface up to isotopy.

> 3   Giving a boundary for Teichmüller space that is natural with respect to the action of diffeomorphisms.

Every closed surface admits a Riemannian metric of constant curvature [Gau65]. Table 1.1 summarizes the possibilities and at the same time establishes a parallel between geometric and the topological properties.

Table 1.1. The three possible geometries on surfaces

Surface	K (curvature)	χ (Euler characteristic)	Remarks
S^2, $\mathbb{R}P^2$	$K = 1$ (Elliptic geometry)	$\chi > 0$	π_1 is finite, $\pi_2 \neq 0$
T^2, K^2 (Torus, Klein bottle)	$K = 0$ (Euclidean geometry)	$\chi = 0$	These are $K(\pi, 1)$'s and their universal covering space is $\mathbb{R}^2$
Genus > 1	$K = -1$ (Hyperbolic geometry)	$\chi < 0$	

Most of Thurston's theorems hold for any compact surface, but in what follows, we restrict ourselves to compact orientable surfaces, possibly with boundary.

1.2   THE SPACE OF SIMPLE CLOSED CURVES

Let M be a compact, connected, orientable surface. We write $\mathcal{S}(M)$ (or, briefly, $\mathcal{S}$) for the set of isotopy classes of simple, closed, connected curves in M that are not homotopic to a point or homotopic to a boundary component of M.

(1) The elements of $\mathcal{S}$ are *not* oriented.

(2) Since two simple closed curves that are homotopic are also isotopic (see [Eps66]), we may replace "isotopy classes" in the above definition with "homotopy classes."

Consider the symmetric map

$$i \colon \mathcal{S} \times \mathcal{S} \to \mathbb{Z}^+ = \{0, 1, 2, \dots\}$$

defined in the following fashion: $i(\alpha, \beta)$ is the minimum number of intersections of a representative for α with a representative for β. This is the *geometric intersection number* (as opposed to the algebraic intersection number).

Example. On the torus T^2, we choose two oriented generators x and y for $\pi_1(T^2)$. Then all elements of $\mathcal{S}$ may be represented by $\gamma(a, b) = ax + by$, where $a, b \in \mathbb{Z}$ and $\gcd(a, b) = 1$; in $\mathcal{S}$, we have $\gamma(a, b) = \gamma(-a, -b)$. The following formula is easy to verify:

$$i\left(\gamma(a, b), \gamma(c, d)\right) = \left| \det \begin{pmatrix} a & b \\ c & d \end{pmatrix} \right|.$$

LEMMA 1.1. *Let M and $\mathcal{S} = \mathcal{S}(M))$ be as above.*
(1) If $\alpha \in \mathcal{S}$, there is a $\beta \in \mathcal{S}$ such that $i(\alpha, \beta) \neq 0$.
(2) If $\alpha_1 \neq \alpha_2$ in $\mathcal{S}$, there is a $\beta \in \mathcal{S}$ such that $i(\alpha_1, \beta) = 0$ and $i(\alpha_2, \beta) \neq 0$.

The proof is given in Exposé 3.

The space of functionals. We consider the set $\mathbb{R}_+^{\mathcal{S}}$ of functions from $\mathcal{S}$ to the nonnegative reals, with the weak topology. The usual multiplication by the positive reals defines rays in $\mathbb{R}_+^{\mathcal{S}}$. The set of these rays is the projective space $P(\mathbb{R}_+^{\mathcal{S}})$; it is given the quotient topology. We have the natural maps

$$\mathcal{S} \xrightarrow{i_*} \mathbb{R}_+^{\mathcal{S}} \setminus 0 \xrightarrow{\pi} P(\mathbb{R}_+^{\mathcal{S}})$$

where the map i_* is defined by $i_*(\alpha)(\beta) = i(\alpha, \beta)$. By statement *(1)* of Lemma 1.1, $i_*(\mathcal{S})$ does not contain 0, and by statement *(2)*, the map $\pi \circ i_*$ is injective.

Consider the completion of $\mathcal{S}$, denoted $\overline{\mathcal{S}}$, which is the closure of $\pi \circ i_*(\mathcal{S})$ in $P(\mathbb{R}_+^{\mathcal{S}})$. The elements of $\overline{\mathcal{S}}$ are represented by sequences $\{(t_n, \alpha_n)\}$, $t_n > 0$, $\alpha_n \in \mathcal{S}$, such that for all β in $\mathcal{S}$, the sequence of real numbers $t_n i(\alpha_n, \beta)$ converges.

Thus, within $P(\mathbb{R}_+^{\mathcal{S}})$, the set $\mathcal{S}$ has a nontrivial topology. Intuitively, we may give a meaning to the notion that "two curves γ, γ' are close to each other." This "proximity" has nothing to do with the respective homotopy classes of the curves, but with the fact that, up to a multiple, in every region of the surface, γ and γ' are more or less made up of the same number of strands, going in more or less the same direction. All of this will be discussed in greater detail in Exposé 4.

We also need to introduce the space $\mathcal{S}'$ of isotopy classes of simple, closed, but not necessarily connected curves in M, whose every component represents an element of $\mathcal{S}$. Two distinct components of the same curve are allowed to be isotopic to each other, so that we may consider scalar multiplication: for an integer $n > 0$ and $\gamma \in \mathcal{S}$, $n\gamma$ is represented by n parallel curves.

As before, we define $i \colon \mathcal{S}' \times \mathcal{S} \to \mathbb{Z}_+$, and obtain the diagram

$$\mathcal{S}' \xrightarrow{\ i_*\ } \mathbb{R}_+^{\mathcal{S}} \setminus 0 \xrightarrow{\ \pi\ } P(\mathbb{R}_+^{\mathcal{S}}).$$

Clearly, i_* respects multiplication by scalars, hence $\pi \circ i_*$ is not injective on $\mathcal{S}'$. But one may easily show that $\pi \circ i_*(\mathcal{S}')$ admits $\overline{\mathcal{S}}$ as its closure (see Exposé 4). In the following, we denote by $\mathbb{R}_+ \times \mathcal{S}$ the cone on $i_*(\mathcal{S})$ in $\mathbb{R}_+^{\mathcal{S}}$. Also, we denote by $M_{g,b}^2$ the surface $\#(T^2) - \bigcup_b D^2$.
$\phantom{by M_{g,b}^2 the surface }{}_g$

THEOREM 1.2. *If M is a closed surface of genus $g \geq 2$, then $\overline{\mathcal{S}}$ is homeomorphic to S^{6g-7} (this is proven in Exposé 4). If $M = M_{g,b}^2$ and $\chi(M) < 0$, then $\overline{\mathcal{S}(M)}$ is homeomorphic to $S^{6g+2b-7}$ (see Exposé 11). Last, $\overline{\mathcal{S}(T^2)} \simeq S^1$ and $\overline{\mathcal{S}(D^2)} = \overline{\mathcal{S}(S^2)} = \overline{\mathcal{S}(S^1 \times [0,1])} = \emptyset$.*

1.3 MEASURED FOLIATIONS

For simplicity, M will be closed. A *measured foliation* on M is a foliation $\mathcal{F}$ with singularities (of the type of a holomorphic quadratic differential $z^{p-2}\, dz^2$, where $p = 3, 4, \dots$) together with a transverse measure that is invariant under holonomy. In the neighborhood of a nonsingular point, there exists a chart $\varphi \colon U \to \mathbb{R}_{x,y}^2$ such that $\varphi^{-1}(y = \text{constant})$ consists of the leaves of $\mathcal{F}|_U$. If $U_i \cap U_j$ is nonempty, there exist transition functions φ_{ij} of the form

$$\varphi_{ij}(x,y) = (h_{ij}(x,y), c_{ij} \pm y)$$

where c_{ij} is a constant. In these charts, the transverse measure is given by $|dy|$.

Remark. The foliations that admit transition functions of the form $(f(x,y), c + y)$ are those that are defined by a closed 1-form ω; away from singularities, y is a local primitive for ω. The singularities of $\mathcal{F}$ are p-saddles ($p \geq 3$) as in Figure 1.1.

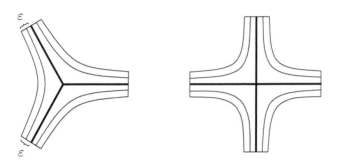

Figure 1.1. p-saddles, for $p = 3$, $p = 4$

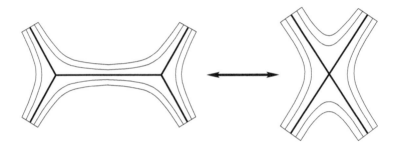

Figure 1.2. Whitehead equivalence

If γ is a simple closed curve in M, we call $\int_\gamma \mathcal{F}$ the *total variation* of the y-coordinate of $p \in \gamma$ as p traverses γ. For $\alpha \in \mathcal{S}$, define

$$I(\mathcal{F}, \alpha) = \inf_{\gamma \in \alpha} \int_\gamma \mathcal{F}.$$

$\mathcal{F}_1$ and $\mathcal{F}_2$ are *Whitehead equivalent* if one may be transformed to the other by isotopies and elementary deformations of the type suggested by Figure 1.2. (Observe that these deformations allow the transverse measure to be carried over.)

Denote by $\mathcal{MF}$ the set of Whitehead equivalence classes. Define

$$I_* \colon \mathcal{MF} \to \mathbb{R}_+^{\mathcal{S}}$$

by

$$I_*(\mathcal{F})(\alpha) = I(\mathcal{F}, \alpha).$$

$\mathcal{F}_1$ and $\mathcal{F}_2$ are *m-equivalent* (or *Schwartz equivalent*) if $I_*(\mathcal{F}_1) = I_*(\mathcal{F}_2)$. Note that Schwartz equivalence is an immediate consequence of Whitehead equivalence.

THEOREM 1.3. *The map I_* injects $\mathcal{MF}$ into $\mathbb{R}_+^{\mathcal{S}}$. What is more, we have $I_*(\mathcal{MF}) \cup 0 = \overline{\mathbb{R}_+ \times \mathcal{S}}$, and if $g > 1$, this set is homeomorphic with $\mathbb{R}^{6g-6}$. In particular, Schwartz equivalence is the same thing as Whitehead equivalence.*

The proof of this theorem is dealt with in Exposés 5 and 6. What is more, since $I_*(\mathcal{MF})$ misses 0, the theorem says that in $P(\mathbb{R}_+^{\mathcal{S}})$ we have $\overline{\mathcal{S}} = \pi \circ I_*(\mathcal{MF})$. This gives a nice geometric representation of the functionals in $\overline{\mathbb{R}_+ \times \mathcal{S}}$.

1.4 TEICHMÜLLER SPACE

We will consider a surface M with $\chi(M) < 0$. Consider the space $\mathcal{H}$ of all metrics on M with constant curvature $K = -1$ such that every component of the boundary of M is a geodesic. Let $\mathrm{Diff}_0(M)$ be the group of diffeomorphisms isotopic to the identity, with the C^∞ topology. As we shall see later, this group acts freely and continuously on $\mathcal{H}$. The orbit space under this action, equipped with the quotient topology, is called the *Teichmüller space* $\mathcal{T}(M)$ (briefly, $\mathcal{T}$). If M is orientable, there is another definition in terms of complex structures on M. The equivalence of the two definitions is a consequence of the uniformization theorem [Wey97].

Remarks. Consider a fixed M, together with another surface $X_\rho = X$ with a hyperbolic metric ρ. If $\varphi \colon M \to X$ is a diffeomorphism, the pair (X, φ) is called a *Teichmüller surface*. Two Teichmüller surfaces (X, φ) and (X', φ') are said to be *equivalent* if there is an isometry $f \colon X \to X'$ such that φ' and $f \circ \varphi$ are isotopic. It is convenient to identify the points of $\mathcal{T}$ with equivalence classes of Teichmüller surfaces.

We also remark here that two diffeomorphisms of M are homotopic if and only if they are isotopic (see [Eps66]).

If M is closed, of genus $g > 1$, a classical theorem of Teichmüller theory asserts that

$$\mathcal{T}(M) \simeq \mathbb{R}^{6g-6},$$

This result, due to Fricke and Klein, will be proven in Exposé 7. Further, we have

$$\mathcal{T}(M_{g,b}^2) \simeq \mathbb{R}^{6g-6+2b}.$$

For all $\theta \in \mathcal{T}$ and $\alpha \in \mathcal{S}$, we define

$$\ell(\theta, \alpha) = \inf_{\gamma \in \alpha}(\theta(\gamma))$$

where $\theta(\gamma)$ denotes the length of γ computed in the metric θ, which is prescribed up to isotopy on M. The metric being fixed, the infimum is attained for a unique geodesic. From the above formula, we obtain the map

$$\ell_* \colon \mathcal{T} \to \mathbb{R}_+^{\mathcal{S}};$$

it can be easily seen that the image of the map misses $I_*(\mathcal{MF}) \cup 0$. The *mapping class group* $\pi_0(\mathrm{Diff}(M))$ acts on Teichmüller space as well as on $\mathcal{S}$, and thus on $\mathbb{R}_+^{\mathcal{S}}$; the map ℓ_* is clearly equivariant.

In Exposé 7, we prove the following theorem.

THEOREM 1.4. *The map ℓ_* is a homeomorphism onto its image.*

It is thus possible to put a natural topology on $\mathcal{T} \cup \overline{\mathcal{S}}$; we consider the topological space $\ell_*(\mathcal{T}) \cup I_*(\mathcal{MF})$, in which the rays in $I_*(\mathcal{MF})$ are identified to points, and we take the quotient topology.

In Exposé 8, we prove the following, in the case where M has no boundary.

THEOREM 1.5. *Let $M = M_{g,b}^2$.*

1. The topological space $\mathcal{T} \cup \overline{\mathcal{S}}$ is homeomorphic to D^{6g-6} if M is closed and $g > 1$; it is homeomorphic to $D^{6g-6+2b}$ if $\chi(M) < 0$.

2. The canonical map $\mathcal{T} \cup \overline{\mathcal{S}} \to P(\mathbb{R}_+^{\mathcal{S}})$ is an embedding.

The space $\mathcal{T} \cup \overline{\mathcal{S}}$, denoted $\overline{\mathcal{T}}$, is the *Thurston compactification* of Teichmüller space. It follows immediately from the definitions that for any diffeomorphism φ of M, the natural action of φ on $\overline{\mathcal{T}}$ is continuous.

If φ is a diffeomorphism of M, and $[\varphi]$ denotes the homeomorphism induced by φ on $\overline{\mathcal{T}}$, then $[\varphi]$ has a fixed point, by the Brouwer fixed point theorem. There are two possibilities.

(i) If $[\varphi]$ has a fixed point in $\mathcal{T}$, then φ is isotopic to an isometry φ' in a hyperbolic metric; in particular, φ' is periodic.

(ii) If $[\varphi]$ fixes a point in $\overline{\mathcal{S}}$, there is a foliation $\mathcal{F}$ such that $\varphi(\mathcal{F})$ is Whitehead equivalent to $\lambda\mathcal{F}$, $\lambda \in \mathbb{R}_+$, where $\lambda\mathcal{F}$ has the same underlying foliation as $\mathcal{F}$, with a transverse measure λ times that for $\mathcal{F}$.

This cursory analysis will be made more precise in what follows.

1.5 PSEUDO-ANOSOV DIFFEOMORPHISMS

We begin with an elementary example. Let $\varphi \in \mathrm{Diff}^+(T^2)$. Up to isotopy, φ is in $\mathrm{SL}(2, \mathbb{Z})$. There are three distinct possibilities for the eigenvalues λ_1 and λ_2 of φ, as follows:

(a) λ_1 and λ_2 are complex ($\lambda_1 = \overline{\lambda_2}$, $\lambda_1 \neq \lambda_2$, $|\lambda_1| = |\lambda_2| = 1$). In this case, φ is of finite order.

(b) $\lambda_1 = \lambda_2 = 1$ (respectively, $\lambda_1 = \lambda_2 = -1$). Up to a change of coordinates,

$$\varphi = \begin{pmatrix} 1 & a \\ 0 & 1 \end{pmatrix} \left[\text{respectively,} \quad \varphi = \begin{pmatrix} -1 & a \\ 0 & -1 \end{pmatrix} \right],$$

which is a *Dehn twist* (respectively, the product of a Dehn twist with the "hyperelliptic involution"). In either case, φ leaves invariant a simple curve.

(c) λ_1 and λ_2 are distinct irrationals. Then φ is an *Anosov* diffeomorphism [Ano69, Sma67].

This analysis is generalized by Thurston to any compact surface:

THEOREM 1.6. *Any diffeomorphism φ on M is isotopic to a map φ' satisfying one of the following three conditions:*

(i) φ' *fixes an element of $\mathcal{T}$ and is of finite order.*

(ii) φ' *is "reducible," in the sense that it preserves a simple curve (representing an element of $\mathcal{S}'$); in this case, one pursues the analysis of φ' by cutting M open along this curve.*

(iii) *There exists $\lambda > 1$ and two transverse measured foliations $\mathcal{F}^s$ and $\mathcal{F}^s$ such that*

$$\varphi'(\mathcal{F}^s) = \frac{1}{\lambda}\mathcal{F}^s$$

and

$$\varphi'(\mathcal{F}^u) = \lambda\mathcal{F}^u.$$

The equalities in *(iii)* mean that the underlying foliations are equal, and the measures are scaled.

Aside from the obvious, saying that $\mathcal{F}^s$ and $\mathcal{F}^u$ are transverse means that their singularities are the same, and that in a neighborhood of the singularities the configuration is similar to that in Figure 1.3. A diffeomorphism that satisfies condition *(iii)* is called *pseudo-Anosov*.

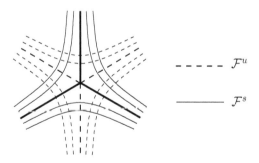

Figure 1.3. Pseudo-Anosov singularities

Theorem 1.6 is proved in Exposé 9. To apply this theorem inductively, we need to extend the theory to the case of surfaces with boundary. This is done in Exposé 11.

In Exposé 12, we show that, for a pseudo-Anosov diffeomorphism φ, the measured foliations $\mathcal{F}^s$ and $\mathcal{F}^u$ represent the only fixed points of $[\varphi]$ in $\overline{\mathcal{T}}$, and two homotopic pseudo-Anosov diffeomorphisms are conjugate by a diffeomorphism isotopic to the identity. The key to these theorems is the following "mixing" property that the pseudo-Anosov diffeomorphism φ possesses: for all $\alpha, \beta \in \mathcal{S}$, we have

$$\lim_{n \to \infty} \frac{i(\varphi^n(\alpha), \beta)}{\lambda^n} = I(\mathcal{F}^s, \alpha) I(\mathcal{F}^u, \beta).$$

Spectral properties of pseudo-Anosov diffeomorphisms. For $\theta \in \mathcal{T}$ and $\alpha \in \mathcal{S}$, we defined in Section 1.4 the positive number $\ell(\theta, \alpha)$. Diffeomorphisms have eigenvalues in the following sense.

THEOREM 1.7. *Let $\varphi \in \mathrm{Diff}(M^2)$. There exists a finite set of algebraic integers $\lambda_1, \ldots, \lambda_k \geq 1$ such that, for each $\alpha \in \mathcal{S}$, there exists λ_j with*

$$\lim_{n \to \infty} \ell(\theta, \varphi^n(\alpha))^{1/n} = \lambda_j$$

for all $\theta \in \mathcal{T}$. Furthermore, φ is pseudo-Anosov if and only if $k = 1$ and $\lambda_1 > 1$; in this case $\lambda_1 = \lambda$ (see Exposés 9 and 11).

Entropy. On a compact metric space X with a continuous map $f \colon X \to X$, we may define the *topological entropy* $h(f)$ (see Exposé 10). If φ is a pseudo-Anosov diffeomorphism, one proves that $h(\varphi) = \log(\lambda)$. Moreover, φ possesses an obvious invariant measure and $h(\varphi)$ is its metric entropy [Sin76].

THEOREM 1.8. *A pseudo-Anosov diffeomorphism minimizes the topological entropy in its isotopy class.*

The list of Thurston's results is much longer, but we end this overview here to get to the heart of the subject.

1.6   THE CASE OF THE TORUS

This case is particularly simple and is treated separately. On the torus T^2, consider the three elements e_1, e_2, e_3 of $\mathcal{S}(T^2)$ shown in Figure 1.4. Let these be oriented for the time being.

Let x_1 and x_2 be the canonical generators e_1 and e_2 with the orientations shown in Figure 1.4.

If γ is a simple oriented curve representing $mx_1 + nx_2$, we find

$$i(e_1, \gamma) = |n|, \quad i(e_2, \gamma) = |m|, \quad i(e_3, \gamma) = |n - m|.$$

These three numbers determine γ in $\mathcal{S}$, but the first two are not sufficient. The three numbers form a degenerate triangle, in the sense that one of them is equal to the sum of the other two.

We now consider the standard simplex with barycentric coordinates X_1, X_2, X_3 (where $X_i \geq 0$, $\sum X_i = 1$). The simplex decomposes into the four regions indicated in Figure 1.5.

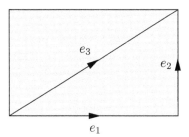

Figure 1.4. The torus T^2

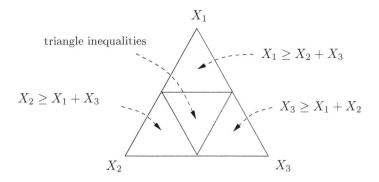

Figure 1.5.

Let $(\nabla \leq)$ be the domain where the triangle inequality holds; the boundary $\partial(\nabla \leq)$ corresponds to degenerate triangles. We think of the standard simplex as being in $\mathbb{R}_+^3$, and we denote by $\mathrm{cone}(\partial(\nabla \leq))$ the cone of half-lines that start at 0 and pass through $\partial(\nabla \leq)$.

To each $\gamma \in \mathcal{S}$, we associate the numbers

$$X_j = \frac{i(e_j, \gamma)}{\sum\limits_{i=1}^{3} i(e_i, \gamma)}, \quad j = 1, 2, 3;$$

a simple calculation shows that we can thus identify $\mathcal{S}$ with the set of rational points of $\partial(\nabla \leq)$.

LEMMA 1.9. *Let $\beta \in \mathcal{S}$. There exists a continuous function*

$$\Phi_\beta \colon \mathrm{cone}(\partial(\nabla \leq)) \to \mathbb{R}_+$$

that is homogeneous of degree 1 (that is, $\Phi_\beta(kv) = k\Phi_\beta(v)$), and that satisfies

$$i(\alpha, \beta) = \Phi_\beta\left(i(\alpha, e_1), i(\alpha, e_2), i(\alpha, e_3)\right)$$

for all $\alpha \in \mathcal{S}$.

Proof. We can give an explicit construction as follows. Suppose that β corresponds to $mx_1 + nx_2$, $n, m \in \mathbb{Z}$, $\gcd(m, n) = 1$. (The only ambiguity is that $-mx_1 - nx_2$ also corresponds to β.) On the surface of the subcone $X_3 = X_1 + X_2$, we set

$$\Phi_\beta(X_1, X_2, X_3) = \left| \det \begin{pmatrix} X_2 & -X_1 \\ m & n \end{pmatrix} \right|$$

On the other two faces, we set

$$\Phi_\beta(X_1, X_2, X_3) = \left| \det \begin{pmatrix} X_2 & X_1 \\ m & n \end{pmatrix} \right|$$

At the intersection of these faces, these formulas agree and Φ_β has the stated property. $\square$

Remark. Φ_β is piecewise linear, a property that we will recover from the other "explicit formulas" of the theory.

Consider now a sequence (λ_n, α_n) with $\lambda_n \in \mathbb{R}_+$ and a sequence $\alpha_n \in \mathcal{S}$ such that, for all $\beta \in \mathcal{S}$, the sequence $\lambda_n i(\alpha_n, \beta)$ converges. Denote by $\lim(\lambda_n, \alpha_n)$ the functional

$$\lim(\lambda_n, \alpha_n)(\beta) = \lim \lambda_n i(\alpha_n, \beta).$$

Since Φ_β is homogeneous, we have

$$\lim(\lambda_n, \alpha_n)(\beta) = \Phi_\beta(\lim(\lambda_n, \alpha_n)(e_1), \lim(\lambda_n, \alpha_n)(e_2), \lim(\lambda_n, \alpha_n)(e_3)).$$

This implies that the bijection of $\mathbb{R}_+ \times \mathcal{S}$, regarded as part of $\mathbb{R}_+^{\mathcal{S}}$, onto the rational rays of $\mathrm{cone}(\partial(\nabla \leq))$ extends to a homogeneous homeomorphism:

$$\overline{\mathbb{R}_+ \times \mathcal{S}} \simeq \mathrm{cone}(\partial(\nabla \leq)) \simeq \mathbb{R}^2.$$

Thus, in $P(\mathbb{R}_+^{\mathcal{S}})$, we have $\overline{\mathcal{S}} \simeq S^1$.

Consider a measured foliation $\mathcal{F}$ of T^2. One can show that $\mathcal{F}$ has no singularities and that it is transversely orientable (this is a consequence of a simple Euler–Poincaré type formula). We can identify $\mathcal{F}$ with a closed nonsingular 1-form. This form is then isotopic to a unique *linear form* (a 1-form with constant coefficients in the canonical coordinates on T^2) [Ste69].

If ω is linear, every curve $\gamma = mx_1 + nx_2$ is transverse to ω, or else is contained in a leaf; thus

$$\left| \int_\gamma \omega \right|^2 = I(\omega, \gamma).$$

The form ω is determined up to sign by $I(\omega, e_1)$, $I(\omega, e_2)$, and $I(\omega, e_3)$. The following lemma is now clear.

LEMMA 1.10. *Let $\mathcal{F}$ be a measured foliation on T^2.*

1. *The numbers $I(\mathcal{F}, e_1)$, $I(\mathcal{F}, e_2)$, and $I(\mathcal{F}, e_3)$ form a degenerate triangle.*

2. *If $\beta \in \mathcal{S}$, we have*

$$I(\mathcal{F}, \beta) = \Phi_\beta(I(\mathcal{F}, e_1), I(\mathcal{F}, e_2), I(\mathcal{F}, e_3))$$

where Φ_β is the function from Lemma 1.9.

The first point is clear from Figure 1.6.

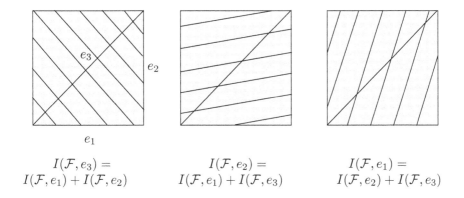

$$I(\mathcal{F}, e_3) = \qquad I(\mathcal{F}, e_2) = \qquad I(\mathcal{F}, e_1) =$$
$$I(\mathcal{F}, e_1) + I(\mathcal{F}, e_2) \qquad I(\mathcal{F}, e_1) + I(\mathcal{F}, e_3) \qquad I(\mathcal{F}, e_2) + I(\mathcal{F}, e_3)$$

Figure 1.6. Proof of Lemma 1.10 part (1)

As a consequence, in $P(\mathbb{R}_+^{\mathcal{S}})$, we have $\pi \circ I_*(\mathcal{MF}) = \overline{\mathcal{S}}$.

In Section 1.4, we defined Teichmüller space for surfaces of negative Euler characteristic. For T^2, one may give an analogous definition by considering flat metrics ($K = 0$) such that $\text{Area}(T^2) = 1$. (This normalization condition is not needed in the hyperbolic case, since, by the Gauss–Bonnet theorem, the area of a surface is determined by its topology.)

Remark 1. Instead of this normalization, one may work with flat metrics up to scaling. What is more, if T^2 is given a complex structure, its universal cover $\widetilde{T^2}$ is isomorphic to $\mathbb{C}$ and the group of linear automorphisms of $\mathbb{C}$, namely $\{z \mapsto \alpha z + \beta : \alpha, \beta \in \mathbb{C}\}$, coincides with the group of orientation-preserving maps of $\mathbb{R}^2$ preserving the Euclidean metric up to a scalar. From this, one easily deduces the equivalence of our definition of $\mathcal{T}$ with the classical definition as the set of complex marked structures on T^2, up to isotopy.

Remark 2. A flat structure on T^2 has an underlying affine structure. If we fix two generators e_1 and e_2 for $\pi_1(T^2)$, the affine structure underlying the metric

ρ is determined by the data of all the geodesics in the class e_i, which are parallel closed curves, as well as all of the numbers

$$\mathrm{dist}\left(\frac{\Delta}{\Delta'}\right) \Big/ \mathrm{dist}\left(\frac{\Delta'}{\Delta''}\right) \in \mathbb{R}_+$$

where $\{\Delta, \Delta', \Delta''\}$ is an arbitrary triple of geodesics all parallel to e_1 or all parallel to e_2. It is easy to see that any affine structures on T^2 with the same data are isotopic to each other. Thus we may always represent an element of $\mathcal{T}$ by a flat metric ρ whose underlying affine structure is the canonical structure (this choice will always be made in what follows). In other words, the usual straight lines are the geodesics for ρ.

To an element $\rho \in \mathcal{T}$, we may associate the triple (X_1, X_2, X_3), where $X_j = \rho(e_j)/\sum_k \rho(e_k)$, and where $\rho(e_j)$ is the length of the geodesic e_j in the metric ρ.

LEMMA 1.11. *The above map is a homeomorphism* $\mathcal{T} \to \mathrm{int}(\nabla \leq)$.

Proof. It is clear that (X_1, X_2, X_3) satisfies the triangle inequality. Let Δ be a triangle in $\mathbb{R}^2$; every assignment of lengths to the sides satisfying the triangle inequality determines a flat metric on $\mathbb{R}^2$ compatible with the affine structure; this is invariant under the group of translations, hence induces a metric on T^2. This shows surjectivity. For injectivity, we note that two flat metrics with standard affine structures giving the same lengths to the sides of Δ are identical. The topology is left for the reader.

In other words, the composition

$$\mathcal{T} \xrightarrow{\ell_*} \mathbb{R}_+^{\mathcal{S}} \xrightarrow{\mathrm{proj}} \mathbb{R}_+^{(e_1, e_2, e_3)}$$

is a homeomorphism of $\mathcal{T}$ onto its image. To see that ℓ_* is a homeomorphism onto its image, note that the length of a given line segment depends continuously on the lengths assigned to e_1, e_2, e_3 (classical trigonometry!).

We have

$$\ell_*(\mathcal{T}) \bigcap I_*(\mathcal{MF}) = \emptyset.$$

Indeed, if ω is a differential form, there exists a sequence γ_n of simple closed curves such that $\int_{\gamma_n} \omega \to 0$; if α_n denotes the class of γ_n in $\mathcal{S}$, we have $I_*(\omega)(\alpha_n) \to 0$, while for a given metric the lengths of the closed geodesics do not approach zero.   □

To prove the analog of Theorem 1.5 for the torus T^2, it remains to prove the following lemma.

LEMMA 1.12. *Let ρ_n be a sequence of flat metrics (normalized to the canonical affine structure), λ_n a sequence of positive reals, and ω a linear form. Suppose that, for $j = 1, 2, 3$, we have*

$$\lambda_n \rho_n(e_j) \to \left| \int_{e_j} \omega \right|.$$

Then, for all closed geodesics α, we have

$$\lambda_n \rho_n(\alpha) \to \left| \int_\alpha \omega \right|.$$

Proof. Let ρ'_n denote the metric $\lambda_n \rho_n$. We treat the case where ω is on the face $X_3 = X_1 + X_2$ of cone$(\partial(\nabla \le))$ (Figure 1.7) and $\int_{e_i} \omega \ne 0$ for $i = 1, 2$.

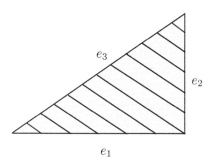

Figure 1.7.

For $j = 1, 2, 3$, we orient e_j so that $\int_{e_j} \omega \ge 0$. Now, let θ_n be the measure of the angle between e_1 and e_2 in the metric ρ'_n.

By the law of cosines, we have

$$[\rho'_n(e_3)]^2 = [\rho'_n(e_1)]^2 + [\rho'_n(e_2)]^2 - 2\rho'_n(e_1)\rho'_n(e_2) \cos \theta_n.$$

The hypothesis then implies that $\cos \theta_n$ tends to -1. If α is a linear segment, say $\alpha = a_1 e_1 + a_2 e_2$, where $a_1, a_2 \in \mathbb{R}$, we have

$$[\rho'_n(\alpha)]^2 = a_1^2[\rho'_n(e_1)]^2 + a_2^2[\rho'_n(e_2)]^2 - 2a_1 a_2 \rho'_n(e_1)\rho'_n(e_2) \cos \theta_n.$$

Thus,

$$[\rho'_n(\alpha)]^2 \to \left[a_1 \int_{e_1} \omega + a_2 \int_{e_2} \omega \right]^2 = \left[\int_\alpha \omega \right]^2.$$

For T^2 the analysis of Theorem 1.6 is trivial. As for Theorem 1.7, it reduces in the case of the torus to a spectral property well-known in linear algebra. $\square$

Exposé Two

Some Reminders about the Theory of Surface Diffeomorphisms

by Valentin Poénaru

2.1 THE SPACE OF HOMOTOPY EQUIVALENCES OF A SURFACE

Let $M = M^2$ be a compact, connected manifold of dimension 2. We will consider the group of diffeomorphisms of M^2, denoted $\mathrm{Diff}(M^2)$. For a subset $A \subset M^2$, we will denote by $\mathrm{HE}(M, A)$ the space of homotopy equivalences $f \colon M \to M$ with $f|_A = \mathrm{id}$; this space is given the topology of uniform convergence.

THEOREM 2.1 (Smale). $\mathrm{Diff}(D^2, \mathrm{rel}\ \partial D^2)$ *is contractible.*

For a proof, see [Cer68], [Sma59].

THEOREM 2.2 ([Cer68]). *The following natural inclusions are homotopy equivalences:*[1]

$$
\begin{array}{ccccc}
O(3) & \hookrightarrow & \mathrm{Diff}(S^2) & \hookrightarrow & \mathrm{HE}(S^2) \\
SO(3) & \hookrightarrow & \mathrm{Diff}(\mathbb{R}P^2) & \hookrightarrow & \mathrm{HE}(\mathbb{R}P^2).
\end{array}
$$

These cases being settled, we may assume that M is a $K(\pi_1(M), 1)$. Choose $* \in M$, let $\mathrm{ev}(*) : \mathrm{HE}(M) \to M$ be the evaluation map, and consider the fibration

$$
\mathrm{HE}(M, *) \xrightarrow{\ \pi\ } \mathrm{HE}(M)
$$
$$
\downarrow {\scriptstyle \mathrm{ev}(*)}
$$
$$
M.
$$

By standard methods from obstruction theory, one proves the following theorem.

THEOREM 2.3. *Let M be a $K(M, 1)$. We have*

$$
\pi_i(\mathrm{HE}(M, *)) = \begin{cases} \mathrm{Aut}(\pi_1(M, *)) & \text{if } i = 0, \\ 0 & \text{if } i > 0. \end{cases}
$$

---

[1] Geoff Mess has informed us that the inclusions $\mathrm{Diff}(S^2) \hookrightarrow \mathrm{HE}(S^2)$ and $\mathrm{Diff}(\mathbb{R}P^2) \hookrightarrow \mathrm{HE}(\mathbb{R}P^2)$ are not homotopy equivalences. Theorem 2.2 is not used in what follows.

Therefore, the homotopy exact sequence of our fibration reduces to

$$1 \to \pi_1(\text{HE}(M)) \to \pi_1(M) \xrightarrow{\partial} \text{Aut}(\pi_1(M)) \to \pi_0(\text{HE}(M)) \to 1.$$

One may easily verify the following facts.

1. If $x \in \pi_1(M)$, then $\partial(x)$ is the inner automorphism corresponding to x.

2. $\pi_1(\text{HE}(M))$ is the center of $\pi_1(M)$. This group is trivial except in the exceptional cases of the torus, where $\pi_1(\text{HE}(T^2)) = \mathbb{Z} \oplus \mathbb{Z}$, and the Klein bottle, where $\pi_1(\text{HE}(K^2)) = \mathbb{Z}$.

3. $\pi_0(\text{HE}(M)) = \text{Aut}(\pi_1(M))/\text{Inn}(\pi_1(M))$, where $\text{Inn}(\pi_1(M))$ is the group of inner automorphisms of $\pi_1(M)$.

2.2   THE BRAID GROUPS

See [Bir74] for more details.

Let X be a topological space and n a positive integer. Let $P_n(X)$ denote $X^n - \Delta$, where Δ is the *big diagonal* of X^n, that is, the set of n-tuples $(x_1, \ldots, x_n)$ of points of X such that for some $i \neq j$, $x_i = x_j$. The symmetric group $\text{Sym}(n)$ acts freely on $P_n(X)$ and we define $B_n(X)$ as $P_n(X)/\text{Sym}(n)$. One thus has a regular covering

$$\text{Sym}(n) \longrightarrow P_n(X)$$
$$\downarrow$$
$$B_n(X).$$

The group $\pi_1(P_n(X))$ is called the *pure braid group* of X on n strands, and $\pi_1(B_n(X))$ is called the *braid group* of X on n strands.

Henceforth, we will take X to be $\mathbb{R}^2$, and we write

$$\begin{aligned}\pi_1(P_n(\mathbb{R}^2)) &= P_n \quad \text{(the pure braid group on n strands)},\\ \pi_1(B_n(\mathbb{R}^2)) &= B_n \quad \text{(the braid group on n strands)}.\end{aligned}$$

We have an obvious exact sequence:

$$1 \to P_n \to B_n \to \text{Sym}(n) \to 1.$$

An element of B_n may be represented in the following manner: fix once and for all a set of n distinct points $x_1, \ldots, x_n$ in the interior of D^2. An element of B_n is a system of arcs in $D^2 \times I$, going from $(x_1, \ldots, x_n) \times 0$ to $(x_1, \ldots, x_n) \times 1$, transverse to every horizontal slice $D^2 \times t$. The arcs do not meet $\partial D^2 \times I$, and everything is defined up to isotopies that leave invariant the boundary of the cylinder and respect the projection $D^2 \times I \to I$.

With this representation, the law of composition in B_n is the same as for cobordisms. The pure braids are those for which the arc leaving $x_i \times 0$ ends at $x_i \times 1$. Figure 2.1 represents an element of B_n.

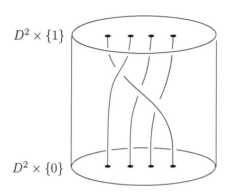

$D^2 \times \{1\}$

$D^2 \times \{0\}$

Figure 2.1. An element of B_n

THEOREM 2.4 (Fadell–Neuwirth). *The map* $P_n(\mathbb{R}^2) \to P_{n-1}(\mathbb{R}^2)$ *that "forgets"* x_n *is a fibration with fiber* $\mathbb{R}^2 - ((n-1)$ *points*$)$.

COROLLARY 2.5. *We have*

$$P_n(\mathbb{R}^2) \simeq K(P_n, 1)$$

and

$$B_n(\mathbb{R}^2) \simeq K(B_n, 1).$$

Remark. The theorem of Fadell–Neuwirth gives us a short exact sequence

$$1 \to F_{n-1} \to P_n \to P_{n-1} \to 1,$$

which is split. Thus, P_n is determined by P_{n-1} and the action of P_{n-1} on the free group F_{n-1}.

We will now give a presentation of the group B_n. In $\mathbb{R}^2$, consider the coordinates (x, y) and, for $p = (p_1, \ldots, p_n) \in B_n(\mathbb{R}^2)$, arrange the indices so that

$$x(p_1) \le x(p_2) \le \cdots \le x(p_n).$$

We define $M_i \subset B_n(\mathbb{R}^2)$ as the set of p such that $x(p_i) = x(p_{i+1})$, as in Figure 2.2.

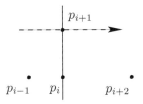

Figure 2.2. Transverse orientation of $M_i - \bigcup_{j \neq i} M_j$

We note the following:

1. $M_i - \bigcup_{j \neq i} M_j$ is a codimension 1 submanifold of $B_n(\mathbb{R}^2)$; it is endowed with a canonical transverse orientation, defined as in Figure 2.2. If the numbering is such that $y(p_{i+1}) > y(p_i)$, a displacement of p_{i+1} along the positive normal pushes p_{i+1} until $x(p_{i+1}) > x(p_i)$.

2. $M_i - \bigcup_{j \neq i} M_j$ is connected.

3. $B_n(\mathbb{R}^2) - \bigcup_i M_i$ is contractible.

These remarks imply that B_n is generated by the simple loops a_i that are based in $B_n(\mathbb{R}^2) - \cup M_i$ and that cross M_i exactly once in the positive direction (without crossing any other stratum). We now find the relations among the a_i by considering what happens in a neighborhood of the stratum of codimension 2 where M_i and M_j meet.

Case 1: $|i - j| \geq 2$. On the level of points in $\mathbb{R}^2$, a point of $M_i \cap M_j$ is a configuration as shown in Figure 2.3. The points p_{i+1} and p_{j+1} may move independently along the dashed horizontal arrows, which gives us a small square that is transverse to $M_i \cap M_j$ in $B_n(\mathbb{R}^2)$, as shown in Figure 2.4. In Figure 2.3, we have drawn the orientations of the transversals to the strata M_i and M_j.

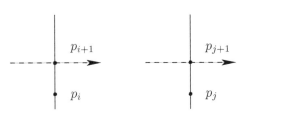

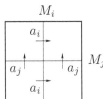

Figure 2.3.   Figure 2.4.

This gives us the relation $a_i a_j = a_j a_i$.

Case 2: $|i - j| = 1$. On the level of points in $\mathbb{R}^2$, we have Figure 2.5, and at the level of $B_n(\mathbb{R}^2)$, we have Figure 2.6. From this we may read off the relation

$$a_i a_{i+1} a_i = a_{i+1} a_i a_{i+1}.$$

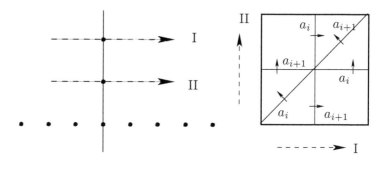

Figure 2.5.   Figure 2.6.

We have thus proven the following theorem.

THEOREM 2.6 (E. Artin). *The braid group B_n admits generators $a_1, a_2, \ldots, a_{n-1}$ and the relations*

$$a_i a_j = a_j a_i \quad \text{for } |i - j| > 1,$$
$$a_i a_{i+1} a_i = a_{i+1} a_i a_{i+1}.$$

COROLLARY 2.7. $B_3 = \pi_1(S^3 - \text{the trefoil knot})$.

[The explanation of this "coincidence" is this: $B_n(\mathbb{R}^2)$ may be identified with the set of complex monic polynomials of degree n, having distinct roots. Thus $B_n = \pi_1(\mathbb{C}^n - \text{the discriminant locus}) \ldots]$

The generator a_i is the following braid:

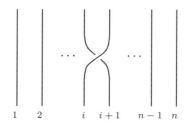

1   2 i $i+1$ $n-1$ n

and the relation $a_i a_{i+1} a_i = a_{i+1} a_i a_{i+1}$ may be visualized as follows:

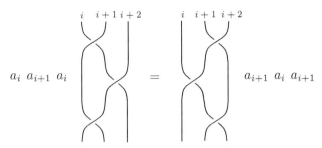

In particular, $B_2 \cong \mathbb{Z}$ and the generator a_1 is

Similarly, $P_2 \cong \mathbb{Z}$ is generated by

and the natural inclusion $P_2 \hookrightarrow B_2$ is multiplication by 2: $\mathbb{Z} \xrightarrow{\times 2} \mathbb{Z}$.

2.3   DIFFEOMORPHISMS OF THE PAIR OF PANTS AND THE SPACES $A(P^2)$, $A'(P^2)$

Let $K \subset \operatorname{int} D^2$ be a finite set of cardinality k. We introduce the following notation:

$$\begin{aligned}
\operatorname{Diff}(D^2, \operatorname{rel}(K, \partial)) &= \{\varphi \in \operatorname{Diff}(D^2) : \varphi|_{K \cup \partial D^2} = \operatorname{id}\}, \\
\operatorname{Diff}(D^2, K, \operatorname{rel} \partial) &= \{\psi \in \operatorname{Diff}(D^2) : \psi(K) = K, \psi|_{\partial D^2} = \operatorname{id}\}.
\end{aligned}$$

We have natural actions of $\operatorname{Diff}(D^2, \operatorname{rel} \partial)$ on $B_k(\operatorname{int} D^2)$ and on $P_k(\operatorname{int} D^2)$, which give us two fibrations:

$$\operatorname{Diff}(D^2, K, \operatorname{rel} \partial) \hookrightarrow \operatorname{Diff}(D^2, \operatorname{rel} \partial) \to B_k(\operatorname{int} D^2)$$

and

$$\operatorname{Diff}(D^2, \operatorname{rel}(K, \partial)) \hookrightarrow \operatorname{Diff}(D^2, \operatorname{rel} \partial) \to P_k(\operatorname{int} D^2).$$

Applying the theorem of Smale that $\mathrm{Diff}(D^2, \mathrm{rel}\,\partial)$ is contractible, we obtain the following corollary.

COROLLARY 2.8. *We have:*

1. *Each connected component of* $\mathrm{Diff}(D^2, \mathrm{rel}(K, \partial))$ *and of* $\mathrm{Diff}(D^2, K, \mathrm{rel}\,\partial)$ *is contractible.*

2. *We have canonical isomorphisms:*

$$P_k \cong \pi_0(\mathrm{Diff}(D^2, \mathrm{rel}(K, \partial))), \tag{2.1}$$
$$B_k \cong \pi_0(\mathrm{Diff}(D^2, K, \mathrm{rel}\,\partial)). \tag{2.2}$$

We will now consider the compact manifold with boundary P^2, which is the disk with two holes, or "*pair of pants*" (see Figure 2.7).

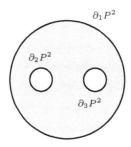

Figure 2.7. The "pair of pants" P^2

Remark. The space

$$\mathrm{Diff}(P^2, \partial_2, \partial_3, \mathrm{rel}\,\partial_1) =$$
$$\{\varphi \in \mathrm{Diff}(P^2) : \varphi|_{\partial_1 P^2} = \mathrm{id}, \varphi(\partial_2 P^2) = \partial_2 P^2, \varphi(\partial_3 P^2) = \partial_3 P^2\}$$

obviously has the same homotopy type as $\mathrm{Diff}(D^2, \mathrm{rel}(K, \partial))$.

PROPOSITION 2.9. $\pi_0(\mathrm{Diff}(P^2, \mathrm{rel}\,\partial)) = \mathbb{Z} \oplus \mathbb{Z} \oplus \mathbb{Z}$.

Proof. Considering the 1-jets of the diffeomorphisms at the two points of K, we have a fibration

$$\mathrm{Diff}(P^2, \mathrm{rel}\,\partial P^2) \longrightarrow \mathrm{Diff}(D^2, \mathrm{rel}(K, \partial))$$
$$\downarrow$$
$$S^1 \times S^1$$

from which we get the exact sequence

$$0 \to \pi_1(S^1 \times S^1) \to \pi_0(\mathrm{Diff}(P^2, \mathrm{rel}\, \partial P^2)) \to P_2 \to 0.$$

One may verify that this sequence splits, that the extension is central, and that the action of P_2 on $\pi_1(S^1 \times S^1)$ is trivial, which gives the stated result.   □

We now consider

$$\mathrm{Diff}^+(P^2, \partial_1, \partial_2, \partial_3) = \{\varphi \in \mathrm{Diff}^+(P^2) : \varphi(\partial_i P^2) = \partial_i P^2\}.$$

PROPOSITION 2.10. $\mathrm{Diff}^+(P^2, \partial_1, \partial_2, \partial_3)$ *is contractible.*

Proof. By restriction of an element $\varphi \in \mathrm{Diff}^+(P^2, \partial_1, \partial_2, \partial_3)$ to $\partial_1 P^2 = \partial D^2$, we have a fibration:

$$\underbrace{\mathrm{Diff}(P^2, \partial_2, \partial_3, \mathrm{rel}\, \partial_1)}_{P_2 = K(\mathbb{Z}, 0)} \hookrightarrow \mathrm{Diff}^+(P^2, \partial_1, \partial_2, \partial_3)$$

$$\downarrow \text{restriction}$$

$$\mathrm{Diff}^+(S^1) = K(\mathbb{Z}, 1).$$

One can check that the map

$$\pi_1(\mathrm{Diff}^+(S^1)) \xrightarrow{\partial} \pi_0(\mathrm{Diff}(P^2, \partial_2, \partial_3, \mathrm{rel}\, \partial_1)) = P_2$$

is an isomorphism, which gives the result.   □

Now let N be a compact surface with (unspecified) nonempty boundary. Define $A(N)$ as the set of isotopy classes of arcs $I \subset N$ that have $\partial I \subset \partial N$ and that represent nontrivial elements of $\pi_1(N, \partial N)$; during an isotopy, each end of an arc is free to move on its respective connected component of ∂N. We define $A'(N)$ similarly, but with several pairwise disjoint arcs.

COROLLARY 2.11. $A(P^2)$ *consists of exactly six elements, classified by the connected components of ∂P^2 in which the endpoints of the respective arcs fall.*

Proof. Let τ and τ' be two representatives of elements of $A(P^2)$ with their endpoints in the same connected component of ∂P^2. We may easily check that there is an orientation preserving diffeomorphism

$$(P^2, \tau) \xrightarrow{\psi} (P^2, \tau').$$

Since $\pi_0(\mathrm{Diff}^+(P^2, \partial_1, \partial_2, \partial_3)) = 0$, this diffeomorphism is isotopic to the identity, which gives the result. The six models are given in Figure 2.8.   □

Now let A' be the set of ordered triples (a_1, a_2, a_3), where $a_i \geq 0$, $a_i \in \mathbb{Z}$, and $\sum a_i \equiv 0 \pmod 2$. If $\tau \in A'(P^2)$, we associate to it

$$i(\tau) = (i(\tau, \partial_1), i(\tau, \partial_2), i(\tau, \partial_3)) \in A',$$

where $i(\tau, \gamma)$ is the number of points τ has in common with γ. For convenience, we adjoin $\emptyset$ to $A'(P^2)$, with $i(\emptyset) = (0, 0, 0)$.

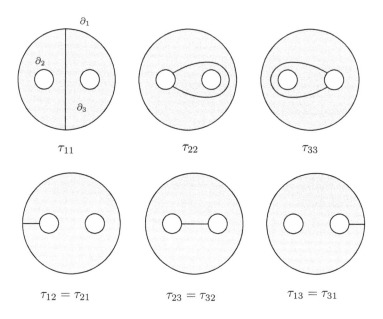

Figure 2.8. The six models for $A(P^2)$

THEOREM 2.12. *The map* $A'(P^2) \xrightarrow{i} A'$ *is a bijection.*

Proof. We begin by constructing a map $A' \xrightarrow{\tau} A'(P^2)$ such that

$$i(\tau(a_1, a_2, a_3)) = (a_1, a_2, a_3).$$

If $(a_1, a_2, a_3) \neq 0$, then the point with barycentric coordinates

$$\left(\frac{a_1}{\sum a_i}, \frac{a_2}{\sum a_i}, \frac{a_3}{\sum a_i} \right)$$

falls in one of the four regions of Figure 2.9.

If (a_1, a_2, a_3) satisfies the triangle inequality, we consider the nonnegative integers

$$x_{12} = \frac{1}{2}(a_1 + a_2 - a_3), \quad x_{23} = \frac{1}{2}(a_2 + a_3 - a_1), \quad x_{31} = \frac{1}{2}(a_3 + a_1 - a_2).$$

We say that an element of $A(P^2)$ is of type τ_{ij} if it connects the ith and jth boundary components of P^2. We define $\tau(a_1, a_2, a_3)$ to be the element of $A'(P^2)$ that consists of $x_{ij} = x_{ji}$ segments of the type τ_{ij}, for $i \neq j$.

If $a_1 \geq a_2 + a_3$, we set

$$x_{11} = \frac{1}{2}(a_1 - a_2 - a_3), \quad x_{12} = a_2, \quad x_{13} = a_3,$$

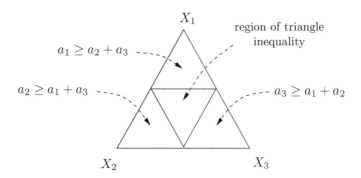

Figure 2.9.

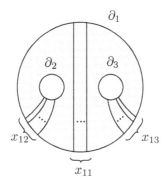

Figure 2.10.

and we define $\tau(a_1, a_2, a_3)$ as in Figure 2.10.

The other cases are treated in a similar manner. One may verify that on $\partial(\nabla \leq)$ the different definitions agree and that $i \circ \tau$ is the identity. Thus i is surjective.

We now remark that the compatible pairs of elements of $A(P^2)$ are exactly those that are joined by a segment in Figure 2.11. The four triangles in Figure 2.11 correspond canonically to the four triangles of Figure 2.9. More precisely, let $x_{\alpha\beta}$ be the number of segments of type $\tau_{\alpha\beta}$ that appear in $\tau \in A'(P^2)$. We have the following four mutually exclusive situations:

1. $x_{\alpha\alpha} = 0$ for $\alpha = 1, 2, 3$, which implies that $i(\tau) \in (\nabla \leq)$.

2. $x_{11} \neq 0$, which implies that $a_1 > a_2 + a_3$.

3. $x_{22} \neq 0$, which implies that $a_2 > a_1 + a_3$.

4. $x_{33} \neq 0$, which implies that $a_3 > a_1 + a_2$.

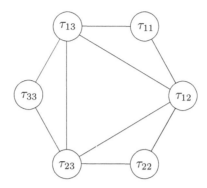

Figure 2.11.

Suppose now that $\tau_1, \tau_2 \in A'(P^2)$ and that $i(\tau_1) = i(\tau_2) \in A'$. We previously deduced that τ_1 and τ_2 are in the same one of the four situations described above; by a linear algebra calculation on the a_1, a_2, a_3 that are (by definition) the same for τ_1 and τ_2, we conclude that the $x_{\alpha\beta}$ are also the same. We still have to prove that if $\tau_1, \tau_2 \in A'(P^2)$ are such that all their $x_{\alpha\beta}$ are equal, then $\tau_1 = \tau_2$. This is already proven if $\sum_{\alpha \leq \beta} x_{\alpha\beta} = 1$. The proof of the general case is an induction on $\sum_{\alpha \leq \beta} x_{\alpha\beta}$. We leave the details to the reader. We have thus proven that i is injective.   □

Remark. Let $\tau \in A'(P^2)$. The group $\pi_0(\mathrm{Diff}^+(P^2, \partial_1, \partial_2, \partial_3, \tau))$ is trivial. In particular, for a given τ, one may not permute the connected components of τ by a diffeomorphism of P^2 that sends each boundary component to itself.

Exposé Three

Review of Hyperbolic Geometry in Dimension 2
and Generalities on Intersection Number

by Valentin Poénaru

3.1 A LITTLE HYPERBOLIC GEOMETRY

Consider a compact surface M that has a Riemannian metric of curvature -1 and whose boundary, if nonempty, is geodesic. The universal cover $\widetilde{M}$ is isometric to a domain in the hyperbolic plane $\mathbb{H}^2$, possibly bounded by geodesics in $\mathbb{H}^2$.

LEMMA 3.1. *If α and β are distinct geodesic arcs in M with the same endpoints, then the closed curve $\alpha \cup \beta$ is not homotopic to a point.*

Proof. If $\alpha \cup \beta$ were homotopic to a point, it would lift to a closed curve in $\widetilde{M}$. But two distinct geodesics in $\mathbb{H}^2$ cannot meet in more than one point. This property of $\mathbb{H}^2$ follows for example from the Gauss–Bonnet formula: for a disk D with a Riemannian metric so that the boundary is a geodesic polygon, we always have

$$\iint_D K = 2\pi - \sum (\text{exterior angles}),$$

where K denotes the curvature. □

LEMMA 3.2. *Let V be a compact Riemannian manifold with totally geodesic boundary. In every (free) homotopy class of maps $S^1 \to V$ there is a geodesic immersion whose length is a lower bound for the length of any loop in its homotopy class.*

Proof. We take a homotopy class $\alpha \in [S^1, V]$, a number $\epsilon > 0$, and an integer N; we set $L = N\epsilon$. We choose ϵ to be smaller than the injectivity radius of the exponential map and N large enough so that α contains at least one curve of length $\leq L$.

Let $I(\alpha, \epsilon, N)$ be the space of continuous maps $S^1 \to V$ in the class α, composed of at most N geodesic arcs of length $\leq \epsilon$ each. This space, with the compact-open topology, is compact, and the length function is continuous. Let φ be a curve that realizes the minimum length in $I(\alpha, \epsilon, N)$. It is easy to check that φ is in fact smooth (if $\partial V \neq \emptyset$, the hypothesis that ∂V is totally geodesic intervenes here).

To see that the length of φ is a lower bound for the class α, it suffices to remark that, if C is a rectifiable curve in α of length $\leq L$, there exists a curve belonging to $I(\alpha, \epsilon, N)$ of length less than or equal to that of C.   □

Remark. Without compactness, with only the hypothesis that the metric is complete, we see that each element of $\pi_1(V, x_0)$ can be realized by a closed geodesic that, in general, is not smooth at x_0.

LEMMA 3.3. *For every nontrivial covering transformation T of $\widetilde{M}$ over M, there exists a unique geodesic invariant under T. It is a lift of the closed smooth geodesic in M that is in the free homotopy class given by the element α of $\pi_1(M, x_0)$ corresponding to T.*

Proof. We give two proofs of existence, and then we prove uniqueness.

Existence. Here is a proof that does not use the curvature assumption. We take as a model for $\widetilde{M}$ the set of continuous paths

$$\{\varphi \colon [0, 1] \to M \mid \varphi(0) = x_0\}$$

subject to the relation of homotopy with endpoints fixed. The projection $p \colon \widetilde{M} \to M$ is given by $\varphi \mapsto \varphi(1)$. The constant path defines the basepoint of $\widetilde{M}$.

Let $\psi \in \widetilde{M}$ with $p(\psi) = y$, and let χ be a path in M such that $\chi(0) = y$. The lift of χ in $\widetilde{M}$ starting from ψ is a one-parameter family of paths in M, obtained by truncating the path $\psi * \chi$; this family begins with ψ and ends with $\psi * \chi$ itself.

The left action of $\pi_1(M, x_0)$ on $\widetilde{M}$ is defined as follows: for $\alpha \in \pi_1(M, x_0)$, which we represent by a loop φ, and for $\psi \in \widetilde{M}$, we set $T_\alpha(\psi) = \varphi * \psi$.

Now, consider the element α for which $T = T_\alpha$. By Lemma 3.2, its free homotopy class contains a smooth closed geodesic g_1. Let x_1 be a point of the image of g_1 and λ a path joining x_0 to x_1; this is chosen so that $\lambda * g_1 * \lambda^{-1}$ belongs to α. If $\widetilde{\lambda * g_1}$ is the lift of $\lambda * g_1$ starting from the basepoint of $\widetilde{M}$, we have

$$\widetilde{\lambda * g_1}(1) = \widetilde{\lambda * g_1 * \lambda^{-1}} * \lambda(1) = T_\alpha(\lambda(1)).$$

Then, if we take in $\widetilde{M}$ the image of $\widetilde{\lambda * g_1}$ and all of its translates by T_{α^n}, $n \in \mathbb{Z}$, we construct a connected component of $p^{-1}(\lambda * g_1)$, consisting of a geodesic g of $\widetilde{M}$ and of segments lifting λ, as in Figure 3.1. By construction, g is invariant under T_α.

We now give a second proof of existence that utilizes the fact that M is a compact surface with a hyperbolic structure. The transformation T_α is an isometry of $\mathbb{H}^2$. As T_α does not have any fixed points, it cannot be elliptic. On the other hand, if T_α were a parabolic isometry of $\mathbb{H}^2$ (having a unique fixed

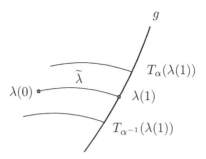

Figure 3.1.

point on the circle at infinity), then for all $\epsilon > 0$ there would be an $x \in \mathbb{H}^2$ such that $d(x, T_\alpha(x)) < \epsilon$. This would imply the existence of closed geodesics of arbitrarily small length in M, which is forbidden by compactness. Thus, T_α is hyperbolic (two fixed points on the circle at infinity). The geodesic g of $\mathbb{H}^2$, which joins them, is hence invariant under T_α, and g/T_α is a smooth closed geodesic in the same free homotopy class as α.

Uniqueness. Let g_1 and g_2 be two distinct geodesics in $\widetilde{M}$, invariant under T. If $g_1 \cap g_2$ is nonempty, the intersection consists of a unique point, which must be invariant under T; but this is impossible.

Hence $g_1 \cap g_2 = \emptyset$. Let $x \in g_1$. We drop a perpendicular from x to g_2, and we denote this geodesic segment by δ. We note that $T\delta \cap \delta = \emptyset$, for otherwise we would have a geodesic triangle where the sum of the (interior) angles is greater than π.

Now g_1, g_2, δ, and $T(\delta)$ form a quadrilateral in which the sum of the interior angles is 2π (see Figure 3.2), but this is impossible by the Gauss–Bonnet formula (or by elementary reasoning).   □

LEMMA 3.4. *Let α be a nontrivial element of $\pi_1(M, x_0)$ There exists a unique smooth closed geodesic in the homotopy class of α.*

Proof. Existence is already ensured by Lemma 3.2. Suppose that g_1 and g_1' are two such geodesics. The "existence" part of the preceding proof provides two distinct geodesics in $\widetilde{M}$, invariant under T_α.

But the "uniqueness" part of the preceding lemma tells us precisely that this is impossible (use the fact that $\pi_1(M, x_0)$ is torsion free).   □

3.2   THE TEICHMÜLLER SPACE OF THE PAIR OF PANTS

The pair of pants P^2, or two-holed disk, is the fundamental "building block" in the theory of surfaces. We recall from Exposé 2 that $\mathrm{Diff}^+(P^2, \partial_1, \partial_2, \partial_3)$ is

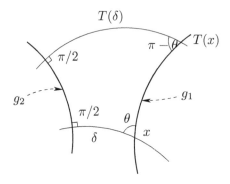

Figure 3.2.

contractible; in particular a diffeomorphism that preserves orientation and sends each boundary component to itself is isotopic to the identity.

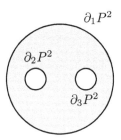

Figure 3.3. The pair of pants P^2

If ρ is a metric of curvature -1 on P^2 for which every boundary component is geodesic, we say that (P^2, ρ) is a P^2-*Teichmüller surface*. By definition, two surfaces (P^2, ρ) and (P^2, ρ') are equivalent if there is a diffeomorphism φ of P^2, isotopic to the identity, such that $\varphi^*\rho = \rho'$. Since $\mathrm{Diff}^+(P^2, \partial_1, \partial_2, \partial_3)$ is connected, the set of equivalence classes—which by definition is the *Teichmüller space* $\mathcal{T}(P^2)$ of P^2—is identified with the quotient of $\mathcal{H}(P^2)$ by $\mathrm{Diff}^+(P^2, \partial_1, \partial_2, \partial_3)$, where $\mathcal{H}(P^2)$ is the space of Riemannian metrics of curvature -1 for which the boundary is geodesic:

$$\mathcal{T}(P^2) = \mathcal{H}(P^2)/\mathrm{Diff}^+(P^2, \partial_1, \partial_2, \partial_3).$$

We endow $\mathcal{H}(P^2)$ with the C^∞ topology and $\mathcal{T}(P^2)$ with the quotient topology. There is a natural continuous map

$$L\colon \mathcal{H}(P^2) \to (\mathbb{R}_+^*)^3 = \{\text{triples of positive numbers}\}$$

defined by
$$L(\rho) = (\ell_\rho(\partial_1 P^2), \ell_\rho(\partial_2 P^2), \ell_\rho(\partial_3 P^2)),$$
where ℓ_ρ denotes the length in the metric ρ. This induces a map that we denote by the same letter:
$$L: \mathcal{T}(P^2) \to (\mathbb{R}_+^*)^3.$$

THEOREM 3.5. *The map* $L: \mathcal{T}(P^2) \to (\mathbb{R}_+^*)^3$ *is a homeomorphism. Moreover,* $L: \mathcal{H}(P^2) \to (\mathbb{R}_+^*)^3$ *admits continuous local sections.*

The classification of P^2-Teichmüller surfaces reduces to the classification of right hyperbolic hexagons, since a hyperbolic pair of pants may be obtained by gluing two isometric hexagons, as shown below (Lemma 3.7). In addition, an "abstract" hyperbolic hexagon X, where every angle is right and where each boundary component is geodesic, is isometric to a hexagon in the hyperbolic plane $\mathbb{H}^2$. To see this, we use X as a fundamental domain, and use symmetries about the sides of X to construct a complete, simply connected hyperbolic manifold Y; by a classical theorem of Cartan–Hadamard [CE75], Y is isometric to $\mathbb{H}^2$. We are therefore interested in the set Hex of (outright) isometry classes of hexagons in $\mathbb{H}^2$, where the angles are right, the sides are geodesic, and one vertex is distinguished. We write $a_1, b_1, a_2, b_2, a_3, b_3$ for the sides, starting from the base vertex and traveling clockwise; see Figure 3.4.

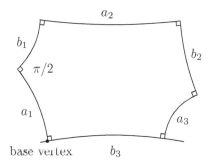

Figure 3.4. Right hyperbolic hexagon with a distinguished vertex

LEMMA 3.6. *The lengths* $\ell(a_1)$, $\ell(a_2)$, *and* $\ell(a_3)$ *establish a bijection from* Hex *to* $(\mathbb{R}_+^*)^3$.

Proof. In turn, we prove the existence and uniqueness of elements Hex corresponding to given elements of $(\mathbb{R}_+^*)^3$.

Existence. Let $\ell_1, \ell_2, \ell_3 > 0$. We want to construct a hexagon X in $\mathbb{H}^2$ such that $\ell(a_i) = \ell_i$ for $i = 1, 2, 3$.

We start by fixing three geodesics G, G', and G'' as in Figure 3.5; G and G'' are a distance ℓ_1 apart. Let $x \in G$ and let L_x be the perpendicular to G starting at x. If x is sufficiently far from x_0, then L_x never meets G'' (we suggest that the reader sketch the picture in the Poincaré model). Let $x(\ell_1)$ be the point of G closest to x_0 that satisfies

$$L_{x(\ell_1)} \cap G'' = \emptyset.$$

We set $f(\ell_1) = d(x_0, x(\ell_1))$.

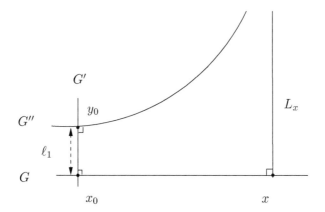

Figure 3.5. Geodesics G, G', G''

We perform the construction in Figure 3.6, which is determined up to isometry by the numbers ℓ_1, ℓ_3, and λ.

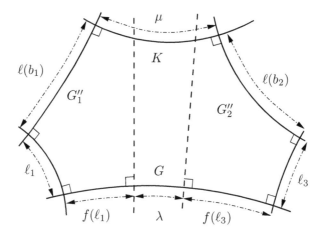

Figure 3.6. Lengths ℓ_1, ℓ_2, and λ determine the hexagon up to isometry

Let $\mu(\lambda)$ be the distance from G_1'' to G_2''; this is a continuous function of the length λ, with $\mu(0) = 0$ and $\mu(+\infty) = +\infty$ (to vary λ, we utilize the fact that there exists a one-parameter group of isometries of $\mathbb{H}^2$ leaving G invariant). As μ takes every positive value, this proves the existence of X.

Uniqueness. As we have just seen, the data of three consecutive sides of a hexagon determines it completely. Thus, if the right hexagons X and X' in Figure 3.7 satisfy $\ell_i = \ell(a_i) = \ell(a_i')$ and are not isometric, then the lengths $\ell(b_3)$ and $\ell(b_3')$ are not equal; say that $\ell(b_3') > \ell(b_3)$.

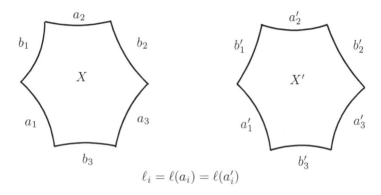

$$\ell_i = \ell(a_i) = \ell(a_i')$$

Figure 3.7.

It is an easy exercise in hyperbolic geometry to see that there exists a (unique) perpendicular from b_3 to a_2 in X. This decomposes the lengths of b_3 and a_2 as shown in Figure 3.8: $\ell(b_3) = \alpha + \beta$; $\ell(a_2) = \gamma + \delta$.

In X', we erect perpendiculars to b_3' at distances α and β from the two endpoints, as in Figure 3.9. In this figure, all of the angles marked by a box are equal to $\pi/2$; the others are not necessarily right.

Figure 3.9 gives a contradiction, since we have $\gamma + \delta > \gamma + \delta$.   □

Remark 1. The uniqueness that we just proved may be interpreted in the following way: if we fix $\ell(a_1)$ and $\ell(a_3)$, the function $\ell(b_3) \to \ell(a_2)$ is monotone; or, the function $\lambda \mapsto \mu(\lambda)$ (Figure 3.6) is a homeomorphism of $\mathbb{R}_+$.

Remark 2. Referring to the notation of Figure 3.4, we may parametrize the set Hex by $(\ell(a_1), \ell(a_2), \ell(a_3))$ or by $(\ell(b_1), \ell(b_2), \ell(b_3))$. The transition from one set of coordinates to the other is by means of a homeomorphism of $(\mathbb{R}_+^*)^3$. [Indeed, we just saw that the transition from $(\ell(a_1), \ell(a_2), \ell(a_3))$ to $(\ell(a_1), \ell(b_1), \ell(a_3))$ is achieved by a homeomorphism of $(\mathbb{R}_+^*)^3$. We may easily verify that the same thing is true for the transition from $(\ell(a_1), \ell(b_3), \ell(a_3))$ to $(\ell(b_3), \ell(a_2), \ell(b_2))$, etc.]

Remark 3. In Figure 3.6, we see that if $\ell_1 = \ell(a_1)$ and $\ell_3 = \ell(a_3)$ are fixed and $\mu = \ell(a_2)$ tends to 0, then $\ell(b_1)$ and $\ell(b_2)$ tend to $+\infty$.

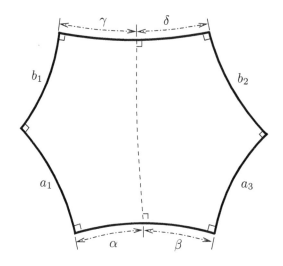

Figure 3.8.

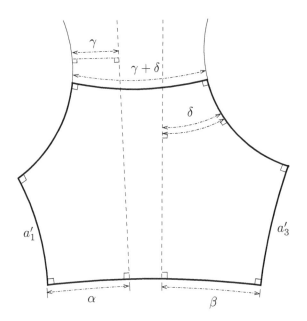

Figure 3.9.

The classification of right-angled hexagons leads to a classification of pairs of pants, because every P^2-Teichmüller surface is the *double* of a hexagon, as indicated precisely in the statement of Lemma 3.7.

LEMMA 3.7. *Suppose a P^2-Teichmüller surface is given.*

1. There exists a unique simple geodesic g_{ij} in P^2 that joins $\partial_i P^2$ to $\partial_j P^2$ and that is perpendicular to both of them. The arcs g_{12}, g_{13}, and g_{23} are mutually disjoint (Figure 3.10).

2. The endpoints of g_{12} and g_{13} cut $\partial_1 P^2$ into segments of equal length (and similarly for $\partial_2 P^2$ and $\partial_3 P^2$).

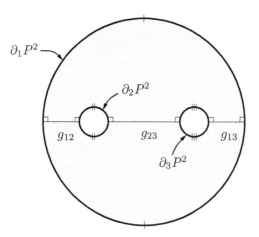

Figure 3.10.

Proof. 1. A path of shortest length joining $\partial_i P^2$ to $\partial_j P^2$ meets the boundary at right angles at its endpoints (apply the first variation formula [CE75]). We deduce right away that it is a simple arc. For uniqueness, we note that the homotopy class is determined by the condition that the path is simple; by an argument using negative curvature as in Lemma 3.3, we obtain statement 1.

2. The arcs g_{12}, g_{13}, and g_{23} cut P^2 into two right hexagons. These are isometric since they have three equal sides.   □

Proof of Theorem 3.5. We proceed in several steps.

1. Surjectivity. Given $\ell_1, \ell_2, \ell_3 > 0$, we may construct a unique right hexagon X with $\ell(a_i) = \ell_i/2$ for $i = 1, 2, 3$ (Lemma 3.6). To form the pair of pants, we take two copies of X and glue them together along b_1, b_2, and b_3. Thus, we have $\ell(\partial_i P^2) = 2\ell(a_i) = \ell_i$, and this gives the surjectivity of L.

2. Uniqueness. Let $\rho', \rho'' \in \mathcal{H}(P^2)$, such that $\ell_i = \ell_{\rho'}(\partial_i P^2) = \ell_{\rho''}(\partial_i P^2)$, for $i = 1, 2, 3$. We are going to prove that there exists $f \in \mathrm{Diff}^+(P^2, \partial_1, \partial_2, \partial_3)$ that takes ρ' to ρ''.

By Lemma 3.7, $(P^2, \rho') = X_1' \cup X_2'$ and $(P^2, \rho'') = X_1'' \cup X_2''$, where X_1', X_2', X_1'', and X_2'' are right-angled hexagons, parametrized by $(\ell_1/2, \ell_2/2, \ell_3/2)$.

Hence, there exist isometries of $X_1' \to X_1''$ and $X_2' \to X_2''$; the desired f is the "union" of these two isometries.

$\textit{3. Continuity.}$ We just proved that the continuous map

$$L \colon \mathcal{T}(P^2) \to (\mathbb{R}_+^*)^3$$

is bijective. To prove that L^{-1} is continuous, it suffices to show that $L \colon \mathcal{H}(P^2) \to (\mathbb{R}_+^*)^3$ admits continuous local sections. It will be more convenient to change coordinates in $(\mathbb{R}_+^*)^3$, changing from the lengths of the boundary curves to the lengths $\ell_{12}, \ell_{23}, \ell_{13}$ of the geodesics g_{12}, g_{23}, g_{13} (Figure 3.10). This gives a new continuous map

$$\Lambda \colon \mathcal{H}(P^2) \to (\mathbb{R}_+^*)^3,$$

and it will suffice to prove that Λ has continuous local sections.

We begin with a few preliminaries. Let E be the portion of $\mathbb{R}^2$ that is the union of

$$E_0 = \{-1 \le y \le 1,\ x = 0\} \quad \text{and} \quad E_1 = \{-1 \le y \le 0,\ 0 \le x \le 1\}.$$

We define $C^\infty(E)$ as the set of functions $f \colon E \to \mathbb{R}$ such that $f|_{E_0} \in C^\infty(E_0)$ and $f|_{E_1} \in C^\infty(E_1)$. We have a natural topology on $C^\infty(E)$ coming from the C^∞ topologies of $C^\infty(E_0)$ and $C^\infty(E_1)$.

LEMMA 3.8. $\textit{There is a continuous map } \epsilon \colon C^\infty(E) \to C^\infty(\mathbb{R}^2) \textit{ such that}$

$$\epsilon(f)|_E = f.$$

$\textit{Proof.}$ Let $f \in C^\infty(E)$. By applying a result of Seeley [See64], we may extend the normal derivative of $f|_{E_0 \cap E_1}$ to all of E_0. This gives us a first extension of $C^\infty(E)$ in the C^∞–Whitney jets on E (we use the fact that E_0 and E_1 are in regular position). Then, we apply the Whitney extension theorem [Mal67]. $\square$

By definition, a $\textit{truncated hexagon}$ is a set consisting of the boundary of a C^∞ hexagon in $\mathbb{R}^2$ and the collar neighborhoods of three alternating sides (Figure 3.11).

Figure 3.11. $Z =$ truncated hexagon

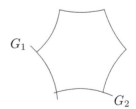

Figure 3.12.

The C^∞ structure of the truncated hexagon Z is locally (where there could be problems) like that of E. From Lemma 3.8 and some classical arguments, we deduce the following lemma.

LEMMA 3.9. *Let* $\mathrm{Emb}(Z, \mathbb{R}^2)$ *be the set of* C^∞ *embeddings of* Z *into* $\mathbb{R}^2$, *with the* C^∞ *topology. If* $\varphi\colon (\mathbb{R}^n, 0) \to \mathrm{Emb}(Z, \mathbb{R}^2)$ *is a germ of a* C^∞ *function, we may lift* φ *to a germ* $\Phi\colon (\mathbb{R}^n, 0) \to \mathrm{Diff}(\mathbb{R}^2)$ *such that* $\Phi(0) = \mathrm{Id}$ *and* $\varphi(t) = \Phi(t)(\varphi(0))$.

Now let $\ell^0 = (\ell_{12}^0, \ell_{23}^0, \ell_{13}^0) \in (\mathbb{R}_+^*)^3$ and let $X(\ell^0)$ be a right hyperbolic hexagon in $\mathbb{H}^2$ parametrized by ℓ^0. Let G_1 and G_2 be two geodesics forming two consecutive sides of $X(\ell^0)$. For ℓ near ℓ^0 in $(\mathbb{R}_+^*)^3$, we consider the hexagon $X(\ell)$ lying on $G_1 \cup G_2$ like $X(\ell^0)$ (Figure 3.12). For each ℓ, the double of $X(\ell)$ along the "marked" sides (those whose lengths are the parameters ℓ_{ij}) is a hyperbolic pair of pants, denoted by $2X(\ell)$.

The problem is to find a diffeomorphism $\overline{\psi}(\ell)\colon 2X(\ell) \to 2X(\ell^0)$, so that the metric $\rho(\ell)$—the image of the natural metric of $2X(\ell)$ under $\overline{\psi}(\ell)$—depends continuously on ℓ as an element of $\mathcal{H}(2X(\ell^0))$.

For small fixed $\epsilon > 0$ (independent of ℓ), we consider in $X(\ell)$ the geodesic collars of radius ϵ along the marked sides; we thus associate to $X(\ell)$ a truncated hexagon $Z(\ell)$. Every rectangle of $Z(\ell)$ is foliated on one hand by the geodesics orthogonal to the sides of the hexagon, and on the other by the trajectories orthogonal to these geodesics. It is easy to construct a germ of a continuous function

$$\varphi\colon ((\mathbb{R}_+^*)^3, \ell^0) \to \mathrm{Emb}(Z(\ell^0), \mathbb{R}^2)$$

such that:

1. $\varphi(\ell^0)$ is the standard embedding.

2. $\varphi(\ell)[Z(\ell^0)] = Z(\ell)$.

3. $\varphi(\ell)$ respects the names of the marked sides and the foliations of the rectangles.

By Lemma 3.9, there exists a germ

$$\psi\colon ((\mathbb{R}_+^*)^3, \ell^0) \to \mathrm{Emb}(X(\ell^0), \mathbb{R}^2)$$

such that $\psi(\ell)|_{Z(\ell^0)} = \varphi(\ell)$. Condition 3 ensures then that $2\psi(\ell)$ is a diffeomorphism of the doubles $2X(\ell^0) \to 2X(\ell)$. On the other hand, the construction ensures that the metric on $X(\ell^0)$ obtained from the natural metric on $X(\ell)$ via $\psi(\ell)$ depends continuously on ℓ. Therefore $\overline{\psi}(\ell) = [2\psi(\ell)]^{-1}$ has all of the required properties. $\qquad\square$

3.3 GENERALITIES ON THE GEOMETRIC INTERSECTION OF SIMPLE CLOSED CURVES

In what follows, M is an orientable surface of genus $g \geq 2$. For practicality, we only explain the case where M is closed; the modifications needed for the case of nonempty boundary are left to the reader. We consider the set $\mathcal{S}$ of isotopy classes of simple closed curves in M that are not homotopic to a point. For $\alpha, \beta \in \mathcal{S}$, we define the *geometric intersection number* $i(\alpha, \beta)$ as the minimal number of intersection points of a representative for α with a representative for β. We are led to a map

$$i_* : \mathcal{S} \to \mathbb{R}_+^{\mathcal{S}}.$$

Throughout this exposé, we shall often use the following theorem due to D. Epstein [Eps66]:

Let $f_0 : S^1 \to M$ be a two-sided embedding (i.e., with trivial normal fiber) that is not the boundary of a disk. If f_1 is an embedding homotopic to f_0, then f_0 and f_1 are isotopic. [With a basepoint, the same thing is true if additionally f_0 is not the boundary of a Möbius band.]

In the same article, one finds the relative version:

Let N be a surface with boundary and say A, B are two embedded arcs with $\partial A = \partial B = A \cap \partial N = B \cap \partial N$. If A and B are homotopic with endpoints fixed, then A and B are isotopic with endpoints fixed.

We will also use the following two facts, which may be found in the same article [Eps66].

If a simple closed curve in a surface is homotopic to a point, then it is the boundary of a disk (this is a consequence of the Jordan–Schönflies theorem).

A two-sided embedding of the circle in a surface cannot be homotopic to a k-fold cover of a two-sided simple curve, for $k > 1$.

PROPOSITION 3.10. *Let α_0' and α_1' be two transverse simple closed curves in M that are not homotopic to a point. We suppose that their isotopy classes α_0 and α_1 are distinct. Then the following conditions are equivalent.*

1. $\mathrm{card}(\alpha_0' \cap \alpha_1') = i(\alpha_0, \alpha_1)$.

2. *No simple closed curve formed from an arc of α_0' and an arc of α_1' is homotopic to a point in M.*

3. *Let $p : \widetilde{M} \to M$ be the universal covering. If $\widetilde{\alpha}_0$ and $\widetilde{\alpha}_1$ are connected components of $p^{-1}(\alpha_0')$ and $p^{-1}(\alpha_1')$, respectively, then we have $\mathrm{card}(\widetilde{\alpha}_0 \cap \widetilde{\alpha}_1) \le 1$.*

4. *There exists a Riemannian metric ρ on M where the curvature is -1 and α_0' and α_1' are geodesics.*

Proof. The reader will notice that the following implications are immediate.

1 ⟹ 2. A simple closed curve γ of $\alpha_0' \cup \alpha_1'$ that is homotopic to a point in M is the boundary of a disk D. Furthermore, γ is the union of an arc of α_0' and an arc of α_1'. Through the disk D, we may perform an isotopy of α_1' that decreases the cardinality of its intersection with α_0'.

3 ⟹ 2 by the theory of covering spaces.

4 ⟹ 2 and *3* by Lemma 3.1. □

LEMMA 3.11. *If* $\operatorname{card}(\alpha_0' \cap \alpha_1') > i(\alpha_0, \alpha_1)$, *there exist two distinct points* q_1 *and* q_2 *of* $\alpha_0' \cap \alpha_1'$ *and two (not necessarily simple) paths* Γ_0 *and* Γ_1 *joining* q_1 *to* q_2, *respectively, on* α_0' *and* α_1', *such that the singular loop* $\Gamma_0 * \Gamma_1^{-1}$ *is homotopic to a point in* M. *Hence* $3 \Rightarrow 1$.

Proof. By hypothesis, there exists a homotopy $h_t \colon S^1 \to M$, for $t \in [0,1]$, such that h_0 parametrizes α_0' and such that $h_1(S^1)$ satisfies

$$\operatorname{card}(h_1(S^1) \cap \alpha_1') < \operatorname{card}(\alpha_0' \cap \alpha_1').$$

We may suppose that the isotopy h_t is in general position with respect to α_1', that is, $h \colon S^1 \times [0,1] \to M$ is transverse to α_1'. Thus $h^{-1}(\alpha_1')$ is a submanifold of dimension 1 transverse to the boundary that consists of four types of connected components, as shown in Figure 3.13.

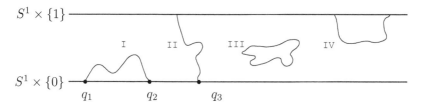

Figure 3.13.

The points $q_1, q_2, q_3, \ldots$ in the figure are exactly the pre-images of $\alpha_0' \cap \alpha_1'$ under the embedding h_0. By assumption, there exists at least one component Γ_1 of type *I*; we obtain Γ_0 by choosing the arc $\overline{q_1 q_2}$ of $S^1 \times \{0\}$ that is homotopic to Γ_1, with endpoints fixed, in $S^1 \times [0,1]$. □

LEMMA 3.12. *Condition 2 ⟹ condition 3.*

Proof. If the components $\widetilde{\alpha}_0$ and $\widetilde{\alpha}_1$ intersect each other in more than one point in $\widetilde{M}$, it is easy to find an embedded disk Δ in $\widetilde{M}$ where the boundary is the union of an arc of $\widetilde{\alpha}_0$ and an arc of $\widetilde{\alpha}_1$. On Δ, we see $p^{-1}(\alpha_0' \cup \alpha_1')$ as in Figure 3.14, where $p^{-1}(\alpha_0')$ is dashed and $p^{-1}(\alpha_1')$ is drawn as a solid line.

We may find a (minimal) disk δ where the boundary is also the union of a dashed arc and a solid arc and where the interior does not meet $p^{-1}(\alpha_0' \cup \alpha_1')$. Because of the minimality, the immersion p embeds the boundary of δ. Now, we may check that p embeds δ: an immersion in codimension 0 that embeds the

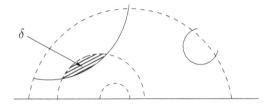

Figure 3.14.

boundary and where the interior does not meet the boundary is an embedding (the number of points of the fiber is locally constant). □

Hence, we have proved the equivalence of conditions *1*, *2*, and *3* of Proposition 3.10.

It remains to prove *1* ⟹ *4*. This follows immediately from Proposition 3.13 and Theorem 3.15 below.

PROPOSITION 3.13. *Let α_0', α_0'', and α_1' be three simple curves in M, each not homotopic to a point. We suppose*

1. *α_0' and α_0'' belong to the same isotopy class α_0, which is* distinct *from the isotopy class α_1 of α_1'; and*

2. *$\mathrm{card}(\alpha_0' \cap \alpha_1') = \mathrm{card}(\alpha_0'' \cap \alpha_1') = i(\alpha_0, \alpha_1)$.*

Then there exists an ambient isotopy of the pair (M, α_1') that pushes α_0' onto α_0''.

Extension. The same proof shows that the proposition remains valid if α_1' is a simple arc representing a nontrivial element of $\pi_1(M, \partial M)$.

Proof of Proposition 3.13. Let $h\colon S^1 \times [0,1] \to M$ be a map that is transverse to α_1', whose restrictions $h|S^1 \times \{0\}$ and $h|S^1 \times \{1\}$ parametrize α_0' and α_0'', respectively.

Claim: The closed components of $h^{-1}(\alpha_1')$ are homotopic to a point in $S^1 \times [0,1]$.

Proof of Claim: Let γ be a closed component of $h^{-1}(\alpha_1')$ that is not homotopic to a point in $S^1 \times [0,1]$. It follows that γ is isotopic to the boundary. Let d be the degree of $h\colon \gamma \to \alpha_1'$. We cannot have $d = 0$, since this would imply that α_0' is homotopic to a point.

We cannot have $|d| > 1$; this would imply that a nontrivial multiple of α_1' is freely homotopic to an embedded curve, namely, α_0'. This is known to be impossible (see the reference to Epstein cited at the beginning of this section). If $|d| = 1$, this means α_0' is homotopic to α_1', which we have excluded.

At this point, we know that the components of $h^{-1}(\alpha_1')$ are of types I, II, III, and IV, as in Figure 3.13. By the second hypothesis, types I and IV do not exist. As $\pi_2(M, \alpha_1') = 0$, it is easy to kill the components of type III. If after this $h^{-1}(\alpha_1')$ is empty, we conclude that α_0' and α_0'' are homotopic—hence isotopic— in $M - \alpha_1'$, and we have the conclusion of the proposition by extending the isotopy to have support in $M - \alpha_1'$. Otherwise, there remain components of type II, which we may deform into vertical segments. However, in general, the resulting homotopy h is singular and does not give an isotopy.

Let $s_1, \ldots, s_n$ be the points of $h^{-1}(\alpha_1') \cap S^1 \times \{0\}$. The s_i cut the circle into intervals $I_1, \ldots, I_n$ and, if $h^{-1}(\alpha_1') = \{s_1\} \times [0,1] \cup \cdots \cup \{s_n\} \times [0,1]$, we may think of $h|_{I_k \times [0,1]}$ as a proper homotopy (i.e., the boundary moves within the boundary) between two embedded arcs of the surface N obtained by cutting M along α_1'. We remark that, by hypothesis 2, $h|_{I_k \times \{0\}}$ represents a nontrivial element of $\pi_1(N, \partial N)$ for all k. Proposition 3.13 is then obtained by applying to each arc Lemma 3.14 below, which generalizes the relative version of the result of Epstein already cited.   □

LEMMA 3.14. *Let N be a surface with boundary, and let γ_0 and γ_1 be two properly embedded arcs in N. Let $h: [0,1] \times [0,1] \to N$ be a proper homotopy between these two arcs, that is, $h(t,0)$ and $h(t,1)$ parametrize γ_0 and γ_1, respectively, and $h(0,u)$ and $h(1,u)$ belong to ∂N for all u.*

The homotopy h is deformable, rel $[0,1] \times \{0,1\}$, to an isotopy from γ_0 to γ_1. Furthermore, if $h(0,u) = h(0,0)$ for all u, or if $h(1,u) = h(1,0)$ for all u, then the deformation may be made through maps with the same properties.

Proof. As usual in these situations, the lemma is clear if γ_0 and γ_1 do not intersect except at their endpoints. Indeed, in this case γ_0 and γ_1 bound a disk in N, through which the required isotopy is done; the isotopy is a deformation of the initial homotopy, since N is an Eilenberg–Mac Lane space.

In the case where they do intersect, we consider the universal covering $p: \tilde{N} \to N$. Consider one component $\tilde{\gamma}_0$ of $p^{-1}(\gamma_0)$ and the union $\tilde{\Gamma}_1$ of all components of $p^{-1}(\gamma_1)$. If we have taken care to begin with an initial isotopy that fixes the endpoints of γ_0 and makes $\operatorname{card}(\gamma_0 \cap \gamma_1)$ as small as possible, then, by the equivalence $1 \iff 3$ of Proposition 3.10, $\tilde{\gamma}_0$ meets every component of $\tilde{\Gamma}_1$ in at most one point.

Let $\tilde{\gamma}_1$ be any component of $\tilde{\Gamma}_1$; we denote by $\tilde{\gamma}_i(0)$ and $\tilde{\gamma}_i(1)$ the endpoints of γ_i. If $\tilde{\gamma}_0$ and $\tilde{\gamma}_1$ meet (somewhere other than at their endpoints), we have the configurations of Figure 3.15. In this figure, the endpoints of the arcs belong to distinct components of $\partial \tilde{N}$, unless explicitly indicated otherwise.

Configuration I cannot occur; indeed, this configuration contradicts the existence of a proper homotopy that separates the two arcs. Similarly, configuration II is excluded in the case where $h(0,u)$ is fixed. By the same argument, configurations III and IV are excluded if, in addition, $h(1,u)$ is fixed. Thus, in the case where the endpoints are fixed, the lemma is totally proven.

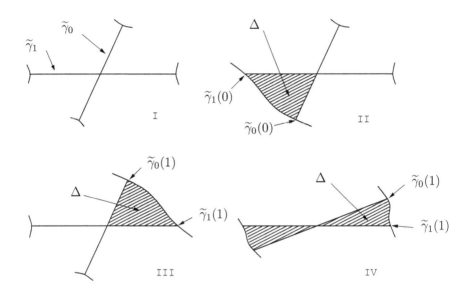

Figure 3.15.

Let us analyze the case where the origin $\tilde{\gamma}_0(0)$ is fixed; then we only have configurations *III* and *IV*. We see in $\tilde{N}$ a triangle Δ. Up to changing components of $\tilde{\gamma}_1$, we may suppose that $(\text{int } \Delta) \cap \tilde{\Gamma}_1 = \emptyset$. Therefore, $p|_\Delta$ is an embedding. There is an isotopy of γ_0 supported in a neighborhood of $p(\Delta)$ that kills at least one point of intersection with γ_1. We continue in this manner until $(\text{int } \tilde{\gamma}_0) \cap \tilde{\Gamma}_1 = \emptyset$. The case where the two endpoints are free is treated similarly.   □

THEOREM 3.15. *Let M be a surface endowed with a metric of curvature -1. Each simple closed curve in M that is not homotopic to a point is isotopic to a simple geodesic. Moreover, two simple geodesics meet in the minimal number of points of intersection in their isotopy classes.*

Proof. The second part of the theorem follows from the implication *3 ⟹ 1* of Proposition 3.10.

Let $f \colon S^1 \to M$ be an embedding that is not homotopic to a point. By Lemma 3.4, f is homotopic to a geodesic immersion g. Let $p \colon \widetilde{M} \to M$ be the universal covering. Let $\tilde{f}_0, \tilde{f}_1 \colon \mathbb{R} \to \widetilde{M}$ be two proper embeddings, with distinct images, lifting f; let $\tilde{g}_0$ and $\tilde{g}_1$ be the geodesic maps to which they are homotopic. By Lemma 3.1, $\tilde{g}_0$ and $\tilde{g}_1$ are embeddings that have at most one point in common. We show that $\tilde{g}_0$ and $\tilde{g}_1$ do not meet.

If $\widetilde{M}$ is regarded as the interior of the Poincaré disk $\mathbb{D}^2$, then $\tilde{g}_i$ has two limit points for $i = 0, 1$. Since the homotopy from $\tilde{g}_i$ to $\tilde{f}_i$ is obtained by lifting a homotopy in M, the hyperbolic distance from $\widetilde{g}_i(x)$ to $\widetilde{f}_i(x)$ is uniformly bounded for $x \in \mathbb{R}$. We know that in a neighborhood of infinity, the Euclidean metric ds^2 is infinitesimally small compared with the hyperbolic ds^2; hence as $x \to \pm\infty$, the Euclidean distance from $\widetilde{g}_i(x)$ to $\widetilde{f}_i(x)$ tends to zero. Thus $\widetilde{f}_i$ has the same limit points on $\partial\mathbb{D}^2$ as $\widetilde{g}_i$. Thus, if $\widetilde{g}_0$ and $\widetilde{g}_1$ have a common point, then by an algebraic intersection argument (or by the Jordan curve theorem), $\widetilde{f}_0$ and $\widetilde{f}_1$ must meet again. This is impossible, since f is an embedding.

Thus we have proved that the image of g is a simple curve covered by g a certain number of times. To see that g is an embedding, we apply the result of Epstein cited at the beginning of the section. □

We can give an application of the theorem that illustrates condition 3 of Proposition 3.10.

COROLLARY 3.16. *Let α_0' and α_1' be two simple closed curves that intersect transversely. We suppose that in the universal cover there are connected components $\widetilde{\alpha}_0$ and $\widetilde{\alpha}_1$ of $p^{-1}(\alpha_0')$ and $p^{-1}(\alpha_1')$, respectively, with $\mathrm{card}(\widetilde{\alpha}_0 \cap \widetilde{\alpha}_1) = \infty$. Then the classes α_0 and α_1 are equal.*

Proof. By the hypothesis of transversality, we have $\mathrm{card}(\alpha_0' \cap \alpha_1') < \infty$. Therefore there are points $* \in \alpha_0' \cap \alpha_1'$ and $x, y \in \widetilde{\alpha}_0 \cap \widetilde{\alpha}_1$, such that $x \neq y$, $p(x) = p(y) = *$. We orient each arc $\widetilde{\alpha}_i$ from x to y and each arc α_i' as $\widetilde{\alpha}_i'$. Consider α_0, α_1 as elements of $\pi_1(M, *)$. The segment from x to y on $\widetilde{\alpha}_0$ (respectively $\widetilde{\alpha}_1$) covers α_0' k times (respectively α_1' l times). We therefore have in $\pi_1(M, *)$ the equality

$$\alpha_0^k = \alpha_1^l.$$

Now, we give M a metric of curvature -1. If g_i denotes the (unique) geodesic of $\widetilde{M}$ invariant under T_{α_i}, we see that $T_{\alpha_0^k} = T_{\alpha_1^l}$ leaves g_0 and g_1 invariant. Thus $g_0 = g_1$, $p(g_0) = p(g_1)$ and α_0', α_1' are (freely) homotopic to the same geodesic in M. □

From the equivalence $1 \iff 2$ of Proposition 3.10, we deduce the following fact. Let α', β', γ' be three simple arcs in M that are each not null-homotopic, and that satisfy $\alpha' \cap \gamma' = \beta' \cap \gamma' = \emptyset$; if $\mathrm{card}(\alpha' \cap \beta')$ is minimal in $M - \gamma'$, then $\mathrm{card}(\alpha' \cap \beta')$ is also minimal in M. This criterion will be used below.

We recall from Exposé 1 that $P(\mathbb{R}_+^{\mathcal{S}})$ is the projective space associated to $\mathbb{R}_+^{\mathcal{S}}$ and that

$$\pi \colon \mathbb{R}_+^{\mathcal{S}} - \{0\} \to P(\mathbb{R}_+^{\mathcal{S}})$$

is the natural projection.

PROPOSITION 3.17. *We have:*

1. *The image of i_* is contained in $\mathbb{R}_+^{\mathcal{S}} - \{0\}$.*

2. *The map $\pi \circ i_*$ (in particular i_*) is injective.*

Proof. It suffices to prove that if $\alpha_1 \neq \alpha_2 \in \mathcal{S}$, there exists $\beta \in \mathcal{S}$ such that

$$i(\alpha_1, \beta) = 0 \neq i(\alpha_2, \beta).$$

If $i(\alpha_1, \alpha_2) \neq 0$, it suffices to take $\beta = \alpha_1$. If $i(\alpha_1, \alpha_2) = 0$, there exist simple curves $\alpha'_1 \in \alpha_1$ and $\alpha'_2 \in \alpha_2$ such that $\alpha'_1 \cap \alpha'_2 = \emptyset$. By cutting m along α'_1, we obtain a surface N containing α'_2 in its interior.

As α'_2 is not isotopic to α'_1, there exists in N a curve β' that cannot be separated from α'_2 in N. If α'_2 does not separate N, we take β' with card($\beta' \cap \alpha'_2$) = 1. If α'_2 separates N into N_1 and N_2, we take $\beta' = I_1 \cup I_2$ where I_j is an arc representing a nontrivial element of $\pi_1(N_j, \alpha'_2)$; this is possible because neither N_1 nor N_2 is an annulus or a disk.

If β is the isotopy class of β' in M, then, by Proposition 3.10, we have $i(\alpha_2, \beta) \neq 0$. $\square$

3.4   SYSTEMS OF SIMPLE CLOSED CURVES AND HYPERBOLIC ISOMETRIES

Consider a system of distinct elements $\alpha_1, \ldots, \alpha_k \in \mathcal{S}$, with the property that $i(\alpha_l, \alpha_q) \leq 1$. We define the complex $\Gamma(\alpha_1, \ldots, \alpha_k)$ having as vertices the $\alpha_1, \ldots, \alpha_k$ and as edges the pairs $\{\alpha_l, \alpha_q\}$ where $i(\alpha_l, \alpha_q) = 1$. We will henceforth suppose that $\Gamma(\alpha_1, \ldots, \alpha_k)$ is a tree.

LEMMA 3.18. *Under the conditions above, let α'_j and α''_j be elements of α_j satisfying* card($\alpha'_l \cap \alpha'_q$) = card($\alpha''_l \cap \alpha''_q$) = $i(\alpha_l, \alpha_q)$. *Then there exists a diffeomorphism of M that is isotopic to the identity and that transforms $\cup \alpha'_j$ into $\cup \alpha''_j$.*

Proof. For $k = 2$, this is Proposition 3.13. For the purposes of induction, assume that $\alpha'_j = \alpha''_j$ for $j \leq l$, $l \geq 2$, the indexing being compatible with the tree structure. Let p, q be such that $p \leq l < q$, and $i(\alpha_p, \alpha_q) = 1$. Let N be the manifold obtained by cutting M along the arcs α'_j, where $j \leq l, j \neq p$. Then α'_p is cut into one or more arcs in N. Let I be one such arc that meets α'_q (α'_q is a closed curve in N since Γ is a tree). As card($\alpha'_q \cap I$) = 1, the arc I represents a nontrivial element of $\pi_1(N, \partial N)$.

We claim that α''_q intersects the same arc I (and not some possibly different component of $\alpha'_p \cap N$). Otherwise, for some $j \neq p$ such that $j \leq l$ and $i(\alpha_j, \alpha_p) = 1$, we have $\alpha_j = \alpha_q$ (look at the pre-image of α'_j in the domain of the homotopy from α'_q to α''_q in M; one of these components is necessarily parallel to the boundary of the annulus). The extension of Proposition 3.13 is now applicable: we have, in N, an isotopy that pushes α''_q onto α'_q and that leaves $\alpha'_p \cap N$ alone. $\square$

Application. Let ρ be a metric of curvature -1 on the surface M. We consider the simple curves $\alpha'_1, \ldots, \alpha'_k$ as in Figure 3.16 (here, we take M closed); note that $M - \cup \alpha'_j$ is a cell. Let α''_j be the geodesic, in the metric ρ, of the isotopy class of α'_j; we verify that $\mathrm{card}(\alpha''_l \cap \alpha''_q) = \mathrm{card}(\alpha'_l \cap \alpha'_q)$. By Lemma 3.18, $M - \cup\alpha''_j$ is a cell. In particular, the configuration of Figure 3.16 can be realized by geodesics.

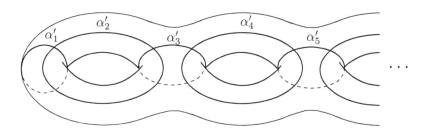

Figure 3.16. The system $\alpha'_1, \ldots, \alpha'_k$ can be realized by geodesics

THEOREM 3.19. *Let ρ be a metric of curvature -1 on a compact surface M. The group $I(M, \rho)$ of isometries of ρ is finite and any isometry isotopic to the identity is the identity.*

Proof. We begin by considering the set M^M of all maps $M \to M$, with the topology of pointwise convergence. By the Tychonov theorem, M^M is compact. We remark, in addition, that on $I(M, \rho)$ the topology of pointwise convergence and the topology of uniform convergence coincide. (Indeed, an isometry is completely characterized by what it does on a sufficiently dense set...) We remark that $I(M, \rho)$ is closed in M^M.

Moreover, we claim that an isometry isotopic to the identity is equal to the identity. Indeed, let φ be such an isometry. The action of φ on $\mathcal{S}$ is trivial. By the uniqueness of geodesics in a given isotopy class $\alpha \in \mathcal{S}$, the geodesic g_α of the class $\alpha \in \mathcal{S}$ is invariant: $\varphi(g_\alpha) = g_\alpha$. We immediately deduce that φ is the identity on the system of geodesics in Figure 3.16. Hence, φ is the identity on the complementary cell.

Thus, $I(M, \rho)$ is discrete. But a closed discrete set in a compact space is finite. $\square$

COROLLARY 3.20. *Let $f \in \mathrm{Diff}(M)$ and let $\mathcal{T}(f)$ be the natural action of f on the Teichmüller space of M (see Exposé 7). If $\mathcal{T}(f)$ has a fixed point, there is a periodic diffeomorphism of M isotopic to f.*

Exposé Four

The Space of Simple Closed Curves in a Surface

by Valentin Poénaru

4.1 THE WEAK TOPOLOGY ON THE SPACE OF SIMPLE CLOSED CURVES

Let M be a closed orientable surface of genus $g \geq 2$. Denote by $\mathcal{S}(M)$ (or more briefly, $\mathcal{S}$) the space of isotopy ($=$ homotopy) classes of unoriented simple closed curves that are not homotopic to a point in M. We have already seen (Section 3.3) that the composite map

$$\mathcal{S} \xrightarrow{i_*} \mathbb{R}_+^{\mathcal{S}} - \{0\} \xrightarrow{\pi} P(\mathbb{R}_+^{\mathcal{S}})$$

is injective. The map i_* extends to a map denoted by the same symbol:

$$i_* \colon \mathbb{R}_+ \times \mathcal{S} \to \mathbb{R}_+^{\mathcal{S}},$$

given by the formula

$$i_*(\lambda, \alpha)(\beta) = \lambda i(\alpha, \beta) \quad \text{for } \lambda \in \mathbb{R}_+ \text{ and } \alpha, \beta \in \mathcal{S}$$

Remark. If $\overline{i_*(\mathbb{R}_+ \times \mathcal{S})}$ denotes the closure of $i_*(\mathbb{R}_+ \times \mathcal{S})$ in $\mathbb{R}_+^{\mathcal{S}}$, we have

$$\pi\left(\overline{i_*(\mathbb{R}_+ \times \mathcal{S})} - \{0\}\right) = \overline{\pi \circ i_*(\mathcal{S})}.$$

This is a general fact about cones.

PROPOSITION 4.1. *In $P(\mathbb{R}_+^{\mathcal{S}})$, the set $\pi \circ i_*(\mathcal{S})$ is precompact.*

For the proof, we begin by choosing on M a metric ρ of curvature -1, and we denote by $\ell(\alpha)$ the ρ-length of the unique geodesic belonging to the class of $\alpha \in \mathcal{S}$.

LEMMA 4.2. *There exists a constant $C = C(M, \rho)$ such that for all $\alpha, \beta \in \mathcal{S}$, we have*

$$i(\alpha, \beta) \leq C\ell(\alpha)\ell(\beta).$$

44

Proof. If $\alpha = \beta$, we have $i(\alpha, \beta) = 0$ and the inequality is clear. Let us suppose therefore that $\alpha \neq \beta$. Let ϵ be a positive number that is smaller than the injectivity radius of the exponential map. The geodesic g_α in the isotopy class α may be covered by fewer than $\ell(\alpha)/\epsilon + 1$ small arcs, each of which is contained in a geodesic disk. The same holds for g_β. By the definition of injectivity radius, a small arc of g_α intersects a small arc of g_β in at most one point. Therefore, in a small arc of g_α, there are at most $\ell(\beta)/\epsilon + 1$ points of intersection with g_β. We therefore find

$$i(\alpha, \beta) = \mathrm{card}(g_\alpha \cap g_\beta) \leq \left(\frac{\ell(\alpha)}{\epsilon} + 1\right)\left(\frac{\ell(\beta)}{\epsilon} + 1\right).$$

As $\ell(\alpha) > \epsilon$, the desired inequality is clear.   □

In M, we now consider the system of elements $\beta_1, \ldots, \beta_{2g+1} \in \mathcal{S}$ represented in Figure 4.1. In Section 3.4, we saw that such a system may be realized by geodesics.

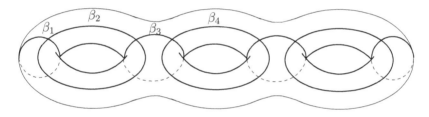

Figure 4.1. The system $\beta_1, \ldots, \beta_{2g+1} \in \mathcal{S}$ may be represented by geodesics

LEMMA 4.3. *There exists a constant c such that for all $\alpha \in \mathcal{S}$,*

$$\sum_j i(\alpha, \beta_j) \geq c\ell(\alpha).$$

Proof. The system $\{g_{\beta_j}\}$ decomposes M into a number of simply connected regions. In each of these, the length of a geodesic arc is bounded, say by L; thus, we have the desired result by taking $c = 1/L$.   □

Proof of Proposition 4.1. For a fixed constant C, consider the subset $S(C) \subset \mathbb{R}_+^{\mathcal{S}}$, defined by

$$S(C) = \left\{ f \in \mathbb{R}_+^{\mathcal{S}} \mid \forall\, \beta \in \mathcal{S},\ f(\beta) \leq C\ell(\beta) \right\}.$$

By the Tychonov theorem, $S(C)$ is compact. Now take C to be the constant from Lemma 4.2, and let $S_0 \subset S(C)$ be the closure in $\mathbb{R}_+^{\mathcal{S}}$ of the set of functionals of the form $i_*(\alpha)/\ell(\alpha)$. By Lemma 4.3, we see that $S_0 \subset \mathbb{R}_+^{\mathcal{S}} - \{0\}$. Moreover, S_0 is compact; thus $\pi(S_0)$ is compact. By Lemma 4.2, we have the inclusion $\pi \circ i_*(\mathcal{S}) \subset \pi(S_0)$; this gives the compactness of $\overline{\pi \circ i_*(\mathcal{S})}$.   □

4.2   THE SPACE OF MULTICURVES

Since $\mathcal{S}$ is difficult to study, we introduce a space that is larger and easier to study. Let $\mathcal{S}'(M)$ (or, briefly, $\mathcal{S}'$) be the space of isotopy classes of closed submanifolds of dimension 1 (not oriented and not necessarily connected) where no component is homotopic to a point. An element of $\mathcal{S}'$ is called a *multicurve*. As in the case of simple curves, we define $i(\alpha, \beta)$ for $\alpha \in \mathcal{S}'$ and $\beta \in \mathcal{S}$, as well as $i_*\colon \mathcal{S}' \to \mathbb{R}_+^{\mathcal{S}}$ and $\pi \circ i_*\colon \mathcal{S}' \to P(\mathbb{R}_+^{\mathcal{S}})$. The minimal intersection between a multicurve and a simple curve is the sum of the minimal intersections with the different components.

Remark.   By the same reasoning as in Section 3.3, we prove that i_* is injective and that two elements α_1 and α_2 of $\mathcal{S}'$ have the same image under $\pi \circ i_*$ if and only if they are integer multiples of the same $\alpha_0 \in \mathcal{S}'$ (there is indeed a natural map $\mathbb{N} \times \mathcal{S}' \to \mathcal{S}'$).

THEOREM 4.4.   *In $P(\mathbb{R}_+^{\mathcal{S}})$, we have*

$$\overline{\pi \circ i_*(\mathcal{S})} = \overline{\pi \circ i_*(\mathcal{S}')}.$$

Applying Theorem 4.4, we obtain the following.

COROLLARY 4.5.   *In $\mathbb{R}_+^{\mathcal{S}}$, we have*

$$i_*(\mathcal{S}') \subset \overline{i_*(\mathbb{R}_+ \times \mathcal{S})}.$$

Proof of Theorem 4.4. It suffices to show that $\pi \circ i_*(\mathcal{S})$ is dense in $\pi \circ i_*(\mathcal{S}')$. Let $\alpha \in \mathcal{S}'$ be represented by a union of pairwise disjoint simple curves $\alpha_1, \alpha_2, \ldots, \alpha_k$. We may choose a simple connected curve γ such that $\mathrm{card}(\gamma \cap \alpha_j)$ is equal to $i(\gamma, \alpha_j)$ and is nonzero for all j. Let $n_1, n_2, \ldots, n_k$ be positive integers. We shall construct an element $\Gamma(n_1, \ldots, n_k)$ of $\mathcal{S}$. Each arc of γ that crosses a small tubular neighborhood of α_j is replaced by an arc with the same endpoints making n_j positive turns (see Figure 4.2, for $n_j = 2$.)

We obtain by this construction a curve $\Gamma(n_1, \ldots, n_k)$ that is well-defined up to isotopy. We prove in Proposition A.1 of Appendix A that for $\beta \in \mathcal{S}$, we have the inequality

$$\left| i(\Gamma(n_1, \ldots, n_k), \beta) - \sum_j n_j i(\gamma, \alpha_j) i(\alpha_j, \beta) \right| \leq i(\gamma, \beta).$$

For any n, set

$$n_j = n \prod_{\ell \neq j} i(\gamma, \alpha_\ell)$$

and denote the resulting curve $\Gamma(n_1, \ldots, n_k)$ by $\Gamma(n)$; we have

$$\left| i(\Gamma(n), \beta) - n \prod_j i(\gamma, \alpha_j) \left[\sum_j i(\alpha_j, \beta) \right] \right| \leq i(\gamma, \beta).$$

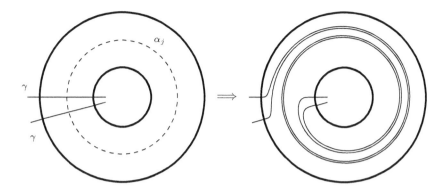

Figure 4.2. The square of a Dehn twist about α_j

In other words, when we projectivize, the contributions of γ to the intersection become negligible as n tends to infinity. Thus the sequence $\pi \circ i_*(\Gamma(n))$ tends to $\pi \circ i_*(\alpha)$. □

Dehn twists. The curve $\Gamma(n_1, \ldots, n_k)$ is alternately described as the image of the curve γ under a diffeomorphism of the surface M. A *Dehn twist* about a curve α in M is a map that acts as a twist on some annular neighborhood of α and acts as the identity outside of this annulus. The isotopy class of the Dehn twist depends only on the isotopy class of α. Also, the direction of the twist depends only on the orientation of the surface, and not on any orientation of α.

Figure 4.2 shows the square of a Dehn twist. The curve $\Gamma(n_1, \ldots, n_k)$ is obtained from γ by product of the n_ith powers of the Dehn twists about the curves α_i.

4.3 AN EXPLICIT PARAMETRIZATION OF THE SPACE OF MULTICURVES

Recall that P^2 denotes the standard pair of pants; the boundary curves are numbered $\partial_1 P^2$, $\partial_2 P^2$, $\partial_3 P^2$. In Section 2.3, we classified the "multi-arcs" of P^2. An element τ of $A'(P^2)$, the space of multi-arcs, is completely characterized by the three integers $m_j = i(\tau, \partial_j P^2)$, $(j = 1, 2, 3)$; a triple of integers that are not all zero describes a multi-arc exactly when $m_1 + m_2 + m_3$ is even.

In each class of $A'(P^2)$, we choose once and for all a representative, which we shall call *canonical*, as designated in Figure 4.3. For each $\tau \in A'(P^2)$ and each $\partial_j P^2$, we choose an arc x_j, a connected component of $\partial_j P^2 - \tau$, as in Figure 4.3. This choice is uniquely defined, since (P^2, τ) does not admit any nontrivial orientation preserving automorphisms.

For each model τ, we chose a *pants seam* $J_1 = J_1(\tau)$ that has the following properties:

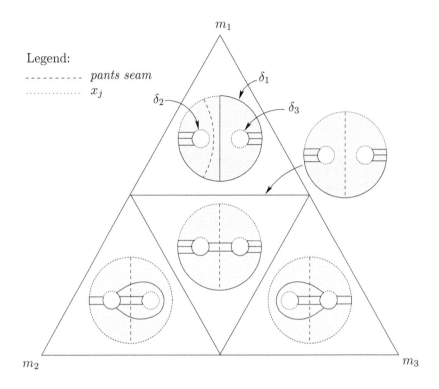

Figure 4.3. Canonical representatives

1. J_1 is a simple arc joining $\partial_1 P^2$ to itself and that cuts P^2 into two regions, one of which contains $\partial_2 P^2$, the other $\partial_3 P^2$.

2. J_1 has one endpoint in the arc $x_1(\tau)$.

3. J_1 has minimal intersection with τ.

Similarly, we construct arcs J_2 and J_3.

Remark. In Exposé 6, we will classify the measured foliations on P^2. The models in Figure 4.3 are the "discrete models" for these foliations, where we see only some of the nonsingular leaves. Further, for the classification of multicurves, we follow a procedure analogous to that which we will follow in the classification of measured foliations, for example, the technique of the pants seam, which is used to recover the way the pairs of pants are glued together to form the surface.

To parametrize $\mathcal{S}'$, we make a number of choices.

(I) We choose $3g - 3$ mutually disjoint simple curves $K_1, K_2, \ldots, K_{3g-3}$ that cut M into $2g - 2$ regions diffeomorphic to pairs of pants. We take these K_i to

have a connected complement in M. It follows that the pairs of pants R_j are embedded in M, that is, each K_i belongs to two distinct pairs of pants.

(II) For each K_j, we choose two simple curves K'_j and K''_j as in Figure 4.4 (this is possible because of the preceding condition). K'_j and K''_j differ by a positive Dehn twist along K_j.

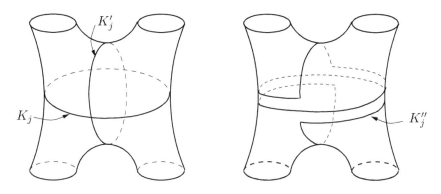

Figure 4.4. K'_j and K''_j differ by a Dehn twist along K_j

(III) We give each K_j a tubular neighborhood $K_j \times [-1, 1]$. These are taken to be pairwise disjoint. The closure of the complement of their union is a number of pairwise disjoint pairs of pants $R'_1, R'_2, \ldots, R'_{2g-2}$.

(IV) Each R'_j is parametrized by P^2, via a diffeomorphism ϕ_j that is fixed (not only up to isotopy).

We consider in $\mathbb{R}_+^{9g-9}$ the cone

$$B = \{(m_i, s_i, t_i) \mid i = 1, \ldots, 3g-3; \ m_i, s_i, t_i \geq 0, \ (m_i, s_i, t_i) \in \partial(\nabla \leq)\}.$$

B is homeomorphic to $\mathbb{R}^{6g-6}$ (the cone on $\partial(\nabla \leq)$ is homeomorphic to $\mathbb{R}^2$). We will construct a "classifying map" $\Phi \colon \mathcal{S}' \to B$.

Let $\beta \in \mathcal{S}'$; we start by defining $m_j(\beta)$ as $i(\beta, K_j)$. These integers determine the model for each pairs of pants R'_k: the corresponding model in P^2 is carried by the diffeomorphism ϕ_k. If the representative β_0 of β is chosen to have minimal intersection with the boundaries of all of the pants R'_k, then $\beta_0 | R'_k$ is isotopic to the model. We therefore choose β_0 equal to the model in all of the pairs of pants R'_k; we say that this representative is in *normal form*. Note that if β_0 has a component isotopic to K_j, this component is contained in the annulus $K_j \times [-1, 1]$.

LEMMA 4.6. *The normal form of β is "unique." Precisely, if β_0 and β_1 are two representatives of β in normal form, then, for all $j = 1, \ldots, 3g-3$, $\beta_0 \cap K_j \times [-1, 1]$ and $\beta_1 \cap K_j \times [-1, 1]$ are isotopic relative to the boundary.*

Proof. We need an extension of Proposition 3.13 to the case that one of the curves is a multicurve. More precisely, the following statement will suffice: if γ_0 is a component of β_0 and if γ_1 is the corresponding component of β_1, then there exists an isotopy of M that pushes γ_0 onto γ_1 and that leaves invariant all of the curves $K_j \times \{-1\}$ and $K_j \times \{1\}$, $j = 1, \ldots, 3g - 3$. Actually, the proof of Proposition 3.13 only needs the following improvement: if $\gamma_0 \cap K_j \times \{\pm 1\} = \gamma_1 \cap K_j \times \{\pm 1\} = \emptyset$, then γ_0 is isotopic to γ_1 in $M - K_j \times \{\pm 1\}$. This assertion is true by the "classical" arguments of Lemma 3.3, except possibly if γ_0 is isotopic to K_j. But then, because of the normal form condition, there is nothing to prove.

Now, in the discussion above, we may replace γ_0 (resp. γ_1) by the collection $\bar{\gamma}_0$ (resp. $\bar{\gamma}_1$) of all components of β_0 (resp. β_1) parallel to γ_0 (resp. γ_1). We may thus construct a normal form β_0' with the following properties:

1. β_0' and β_0 are isotopic by an isotopy that respects the curves $K_j \times \{\pm 1\}$.

2. The collection $\bar{\gamma}_0'$, corresponding to $\bar{\gamma}_0$, coincides with $\bar{\gamma}_1$.

Now let δ_0 be a curve of $\beta_0' - \bar{\gamma}_1$ and let δ_1 be the corresponding curve of $\beta_1 - \bar{\gamma}_1$. Since δ_0 is not parallel to $\bar{\gamma}_1$, we have that δ_0 and δ_1 are isotopic in $M - \bar{\gamma}_1$. Again by the same arguments, we then find that there exists an isotopy of M that is constant on $\bar{\gamma}_1$, that respects the curves $K_j \times \{\pm 1\}$, and that pushes δ_0 onto δ_1. We continue in this way with the rest. In the end, β_0 and β_1 are isotopic by an isotopy that respects all of the curves $K_j \times \{\pm 1\}$.

Now, we claim that the above isotopy may be chosen to be constant in all of the small pairs of pants R_j'. This is clear if $\beta_0 \cap R_j'$ is empty; otherwise, it follows from the fact that $\mathrm{Diff}(P^2, \partial_1, \partial_2, \partial_3)$ is simply connected (Proposition 2.10). This completes the proof. □

The above lemma will be essential for the classification.

The models $\beta_0 \cap R_\ell'$ are equipped with their pants seams. Consider a curve K_j and the two adjacent pairs of pants R_1 and R_2. In the small pairs of pants R_1' and R_2', we have the two pants seams J_1 and J_2 emanating from the respective boundaries parallel to K_j. In $K_j \times [-1, 1]$ there are simple arcs $S_j, S_j', T_j,$ and T_j' such that $J_1 \cup S_j \cup J_2 \cup S_j'$ is isotopic to K_j' and $J_1 \cup T_j \cup J_2 \cup T_j'$ is isotopic to K_j''. If we impose the condition that $\partial S_j = \partial T_j$ and $\partial S_j' = \partial T_j'$, $S_j \cap S_j' = \emptyset$, $T_j \cap T_j' = \emptyset$, then $S_j \cup S_j'$ (resp. $T_j \cup T_j'$) is unique up to isotopy relative to the boundary. Moreover, $T_j \cup T_j'$ is obtained from $S_j \cup S_j'$ by a positive Dehn twist about K_j.

Since the endpoints of these arcs are not in β_0, there is a canonical way to put the arcs into minimal position with β_0. Once this is done, we set

$$s_j(\beta) = \mathrm{card}(\beta_0 \cap S_j)$$

and

$$t_j(\beta) = \mathrm{card}(\beta_0 \cap T_j).$$

LEMMA 4.7. *For each j, the triple $(m_j(\beta), s_j(\beta), t_j(\beta))$ belongs to the boundary $\partial(\nabla \leq)$ of the triangle inequality.*

One should compare this lemma with the classification theorem for $\mathcal{S}'(T^2)$ in Exposé 1.

Proof. The proof is given by Figure 4.5.   □

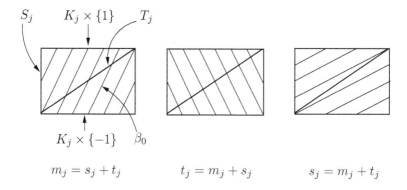

$$m_j = s_j + t_j \qquad t_j = m_j + s_j \qquad s_j = m_j + t_j$$

Figure 4.5. The annulus $K_j \times [-1, 1]$ is cut along S_j

Let $B_0 \subset B$ be the set of nonzero points with integer coordinates that satisfy the following condition: if K_{j_1}, K_{j_2}, K_{j_3}, are on the boundary of the same pair of pants, then $m_{j_1} + m_{j_2} + m_{j_3}$ is even.

THEOREM 4.8. *The map $\Phi \colon \mathcal{S}' \to B$ is a bijection of $\mathcal{S}'$ onto B_0.*

Remark. By an analogous procedure, we will classify measured foliations and Teichmüller structures. Actually, as we will explain, Theorem 4.8 is strictly contained within the classification theorem for measured foliations. But the simplicity of the means implemented here makes it worth including this particular case (for foliations, uniqueness of the normal form is obtained only after a long, roundabout argument).

Proof of Theorem 4.8. The image is obviously contained in B_0. On the other hand, we have a recipe for making a multicurve β from the element $\{m_j, s_j, t_j \mid j = 1, \ldots, 3g - 3\}$ of B_0. By Theorem 2.12, the coefficients m_j determine the arcs in the small pairs of pants R'_k. With these come the pants seams and hence, for each j, we get the arcs S_j and T_j in the annulus $K_j \times [-1, 1]$.

If $m_j = 0$, then $s_j = t_j$ indicates the number of curves of β parallel to K_j. If $m_j \neq 0$, we already have m_j points on $K_j \times \{-1\}$ and on $K_j \times \{1\}$; the coefficients s_j and t_j completely determine the way in which these are joined. It remains to verify that the multicurve constructed in this way has minimal intersection with each K_j, i.e., that $i(\beta, K_j) = m_j$; for this we use the criterion of Proposition 3.10.

As soon as S_j and T_j are fixed, $\beta_0 \cap K_j \times [-1, 1]$ is determined up to isotopy relative to the boundary by s_j and t_j. The injectivity of Φ follows.   □

Remark.   The members of the seminar do not know how to detect which coefficients give a simple connected curve.

Obviously, Φ is homogeneous (of degree 1) with respect to multiplication by an integer scalar. We may thus extend Φ by homogeneity to $\Phi \colon \mathbb{R}_+ \times \mathcal{S} \to B$.

COROLLARY 4.9.   *The map* $\Phi \colon \mathbb{R}_+^* \times \mathcal{S} \to B$ *is injective.*

Proof. If not, there exists $\alpha_0, \alpha_1 \in \mathcal{S}$ and a scalar $\lambda > 0$ such that $\Phi(\alpha_0) = \lambda\Phi(\alpha_1)$. It is easy to see that λ is rational. Thus, we have integers n_0 and n_1 such that $\Phi(n_0\alpha_0) = \Phi(n_1\alpha_1)$. By Theorem 4.8, we have $n_0\alpha_0 = n_1\alpha_1$. It follows immediately that $\alpha_0 = \alpha_1$.   □

Problem. Show directly that $\Phi(\mathbb{R}_+ \times \mathcal{S})$ is dense in B. This is plausible since the (positive) cone on B_0 is dense in B. Of course, the density of $\Phi(\mathbb{R}_+ \times \mathcal{S})$ in B is a consequence of the following theorem which is the "discrete" version of the theorem on foliations, and which will not be proved until Exposé 6.

THEOREM 4.10.   *There exists a closed cone $\mathcal{C}$ in $\mathbb{R}_+^{\mathcal{S}}$ and a continuous map $\theta_C \colon \mathcal{C} \to B$ that is positively homogeneous of degree 1 and that makes the following diagram commute:*

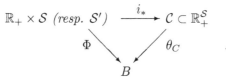

Furthermore, θ_C induces a homeomorphism of $\overline{i_*(\mathbb{R}_+ \times \mathcal{S}')}$ onto B.

Consequences. Theorem 4.10 implies the following:

1. $\Phi(\mathbb{R}_+ \times \mathcal{S})$ is dense in B. (Use Theorem 4.4 and the fact that $\Phi(\mathcal{S}')$ is a "net" by the continuity and the homogeneity of θ_C.)

2. The space $\overline{\pi \circ i_*(\mathcal{S})}$ is homeomorphic to S^{6g-7}.

Remark. The existence of θ_C implies that the coefficients $s_j(\beta)$ and $t_j(\beta)$ are given by continuous, homogeneous, degree 1 functions of the $i(\beta, \alpha), \alpha \in \mathcal{S}$. We will give these explicit formulas in the framework of measured foliations; they make it possible to interpret continuous values of the variables.

Moreover, as Φ is injective for all $\alpha \in \mathcal{S}$, there exists a map $\psi_\alpha \colon B_0 \to \mathbb{N}$ such that for all $\beta \in \mathcal{S}'$, we have

$$i(\beta, \alpha) = \psi_\alpha(\Phi(\beta)).$$

It seems very difficult to make these last formulas explicit.

Appendix A

Pair of Pants Decompositions of a Surface

by Albert Fathi

First, we will give a proof of the inequality used to prove Theorem 4.4. Then, we will apply this inequality to the case where an entire pants decomposition of the surface M gets twisted, instead of a single curve.

Let $\alpha_0, \ldots, \alpha_k$ be a system of mutually disjoint simple closed curves on M. Also, let γ be a simple closed curve whose intersection with each α_j is minimal (among curves isotopic to γ). Let $\{n_j\}$ be a set of positive integers. We construct a curve Γ by acting on γ by the n_jth power of a positive Dehn twist along α_j, for $j = 0, \ldots, k$ (again, the notion of positive twist only depends on the orientation of M). Below, we use the notation $[\ ]$ to mean "isotopy class of."

PROPOSITION A.1. *For each simple curve β, we have the formula*

$$\left| i([\Gamma], [\beta]) - \sum_j n_j i([\gamma], [\alpha_j]) i([\alpha_j], [\beta]) \right| \le i([\gamma], [\beta]).$$

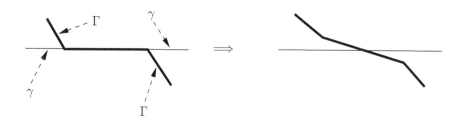

Figure A.1.

Proof. The curve Γ coincides with γ outside of tubular neighborhoods of the α_j. The position of Γ and γ at the endpoints of a common arc is given in Figure A.1. Thus Γ is approximated by a curve denoted by Γ' that crosses each interval of $\Gamma \cap \gamma$ once. This is due to the fact that all of the Dehn twists are positive. By the criterion of Proposition 3.10, we check that $\mathrm{card}(\gamma \cap \Gamma') = i([\gamma], [\Gamma])$.

We observe that $\gamma \cup \Gamma'$ is the image of a continuous map, defined on the disjoint union of $\sum n_j i([\gamma], [\alpha_j])$ copies of S^1, with $n_j i([\gamma], [\alpha_j])$ copies of S^1

going to the free homotopy class of $[\alpha_j]$. Thus, we have the inequality

$$\mathrm{card}(\beta \cap (\gamma \cup \Gamma')) \geq \sum_j n_j i([\gamma], [\alpha_j]) i([\alpha_j], [\beta]).$$

If β does not pass through the points of intersection of γ with Γ', we have

$$\mathrm{card}(\beta \cap (\gamma \cup \Gamma')) = \mathrm{card}(\beta \cap \gamma) + \mathrm{card}(\beta \cap \Gamma').$$

If we take for β a geodesic in some metric of curvature -1 for which γ and Γ' are geodesics (such a metric exists by Proposition 3.10), we have

$$\mathrm{card}(\beta \cap (\gamma \cup \Gamma')) = i([\Gamma], [\beta]) + i([\gamma], [\beta]),$$

which gives one of the desired inequalities.

It remains to prove that

$$i([\Gamma], [\beta]) \leq \sum_j n_j i([\gamma], [\alpha_j]) i([\alpha_j], [\beta]) + i([\gamma], [\beta]).$$

Here, we use the representative Γ rather than Γ'. We choose β to be in minimal position with respect to the α_j and to not pass through the points of intersection of γ with α_j. Each time that β intersects α_j, the curve β crosses the corresponding tubular neighborhood. It thus gives $n_j i([\gamma], [\alpha_j])$ points of intersection with Γ. We therefore have

$$\mathrm{card}(\Gamma \cap \beta) = \mathrm{card}(\beta \cap \gamma) + \sum_j n_j i([\gamma], [\alpha_j]) i([\alpha_j], [\beta]).$$

If, additionally, β has minimal intersection with γ, we have

$$\mathrm{card}(\beta \cap \gamma) = i([\gamma], [\beta]);$$

the left side is always greater than or equal to $i([\Gamma], [\beta])$. $\square$

Let M be a closed surface of genus $g \geq 2$. Let $\mathcal{K} = \{K_1, \ldots, K_{3g-3}\}$ be a system of mutually disjoint simple closed curves on M with the following properties:

1. K_j has connected complement in M;

2. If one cuts M along these curves, one obtains $2g - 2$ pairs of pants (disks with two holes).

We can easily construct a simple curve α that intersects every K_j in an essential way: $i([\alpha], [K_j]) \neq 0$. Let φ be a Dehn twist about α. We set

$$K_j' = \varphi(K_j).$$

Clearly, the system $\mathcal{K}' = \{K_1', \ldots, K_{3g-3}'\}$ has properties 1 and 2.

PROPOSITION A.2. *For all j, k, we have*

$$i([K_j], [K'_k]) \neq 0.$$

Proof. From the inequality of Proposition A.1, it follows that

$$\left| i([K'_k], [K_j]) - i([K_k], [\alpha]) i([\alpha], [K_j]) \right| \leq i([K_k], [K_j]) = 0.$$

$\square$

Remark. We may take α with $i([\alpha], [K_j]) = 2$ for all j. We then obtain $i([K'_k], [K_j]) = 4$ for all j, k.

Exposé Five

Measured Foliations

by Albert Fathi and François Laudenbach

5.1 MEASURED FOLIATIONS AND THE EULER–POINCARÉ FORMULA

Let M be a surface[1] and $\mathcal{F}$ a foliation of M with isolated singularities. By a *transverse invariant measure* we mean a measure μ that is defined on each arc transverse to the foliation and that satisfies the following invariance property:

If $\alpha, \beta \colon [0,1] \to M$ are two arcs that are transverse to $\mathcal{F}$ and that are isotopic through transverse arcs whose endpoints remain in the same leaf, then $\mu(\alpha) = \mu(\beta)$.

If the arc passes through a singularity, the transversality pertains to all points of the arc belonging to a regular leaf.

N.B. In what follows, we restrict ourselves to the case where the measure is regular with respect to Lebesgue measure: every regular point admits a smooth chart (x, y) where the foliation is defined by dy and the measure on each transverse arc is induced by $|dy|$.

Permissible singularities in the interior. For each integer $k > 1$, we consider the singularity of the holomorphic quadratic form $z^k \, dz^2$. We consider

$$\operatorname{Im} \sqrt{z^k \, dz^2} = r^{k/2} \left(r \cos \left(\frac{2+k}{2} \theta \right) d\theta + \sin \left(\frac{2+k}{2} \theta \right) dr \right),$$

which is a form of degree 1, and is well-defined up to sign. It thus defines a measured foliation where the origin is an isolated singularity, and the separatrices are the half-lines given by $r \geq 0$ and $\frac{2+k}{2}\theta = 0 \mod \pi$.

As a model for the singularity we choose a compact domain that contains the origin and that is bounded by arcs transverse to the foliation (*faces*) and arcs contained in the leaves of $\mathcal{F}$ (*sides*).

Remark. Let ω be a closed differential form of degree 1 on M (where $\partial M = \emptyset$) whose singularities are "Morse" (a genericity property). Suppose in addition

---

[1] The theory can be done for nonorientable surfaces. For simplicity, we assume M to be orientable.

that ω does not have a center (critical point of index 0 or 2); then ω defines a measured foliation. It is easy to see that a measured foliation is defined by a closed form if and only if it is transversely orientable in the complement of the singularities.

Permissible singularities on the boundary. The regular points of the boundary are those where the boundary is transverse to the foliation as well as those that have a neighborhood where the boundary is a leaf.

A singular point has a chart that is the image of one of the aforementioned models in the upper half-plane if k is even, or the half-plane of negative real part if k is odd.

Finally, in this entire work, given a measured foliation $(\mathcal{F}, \mu)$ on the manifold M, each point of M has a neighborhood that is the domain of a chart that is foliated isomorphically to one of the models of Figure 5.1.

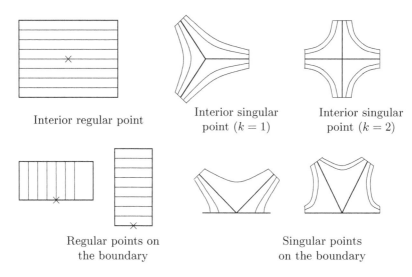

Figure 5.1. Two types of singularities

N.B. A convenient fact is that, by definition, in the chart of a singular point, the separatrices belong to different plaques (a *plaque* is a horizontal line of a foliation in a chart; if it contains a singularity, then it is at an endpoint). Therefore, in M, all leaves are diffeomorphic to intervals of $\mathbb{R}$ or to S^1.

The Euler–Poincaré formula. On the boundary, there are two types of singular-ities for $\mathcal{F}$, as shown on the left-hand side of Figure 5.2. Say that a singularity is of *type (A)* if it lies on a boundary component that is transverse to $\mathcal{F}$ (top of the figure) and of *type (B)* if it lies on a boundary component that is a union of leaves and singular points (bottom of the figure).

To each singularity s, we associate an integer P_s:

$$
\begin{aligned}
P_s \; &= \text{number of separatrices} & &\text{if } s \in \text{int } M \text{ or} \\
& & &\text{if } s \in \partial M \text{ is of type } (B), \\
&= \text{number of separatrices} + 1 & &\text{if } s \in \partial M \text{ is of type } (A).
\end{aligned}
$$

PROPOSITION 5.1 (Euler–Poincaré formula). *Let M be a compact surface, with $\mathcal{F}$ and $\{P_s\}$ as above. We have*

$$
2\,\chi(M) = \sum_{\text{sing } \mathcal{F}} (2 - P_s).
$$

Proof. We begin by reducing to the case where ∂M does not contain singularities, by following the procedure shown in Figure 5.2. By pushing each singularity of the boundary into the interior as shown, we preserve the integer P_s.

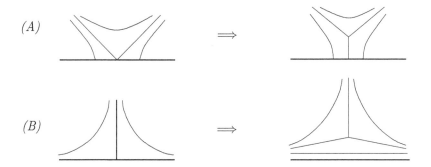

Figure 5.2. Pushing boundary singularities into the interior

Denote by Σ' the set of singular points with an odd number of separatrices, and by Σ'' the set of singular points whose number of separatrices is even. Let $\Sigma = \Sigma' \cup \Sigma''$. We have an orientation homomorphism of the tangent bundle of $\mathcal{F}$:

$$
\pi_1(M - \Sigma) \to \mathbb{Z}/2\mathbb{Z}.
$$

This defines a two-sheeted covering that extends over Σ'' and is branched over Σ'. We therefore have a branched covering $p \colon \widetilde{M} \to M$. The cover $\widetilde{M}$ is equipped with a singular orientable foliation $\widetilde{\mathcal{F}}$, which we may think of as being generated by a vector field $\widetilde{X}$. If s is a singularity of $\widetilde{\mathcal{F}}$, then P_s is an even integer and the index of $\widetilde{X}$ at s is $-\dfrac{P_s}{2} + 1$. Since there are no singularities on the boundary, we have

$$
\chi(\widetilde{M}) = \sum_{\text{sing } \widetilde{X}} \text{index} = \sum_{\text{sing } \widetilde{\mathcal{F}}} \left(-\frac{P_s}{2} + 1 \right),
$$

or

$$\chi(\widetilde{M}) = 2\chi(M) - \operatorname{card} \Sigma' \quad \text{and} \quad \sum_{\operatorname{sing} \widetilde{\mathcal{F}}} 1 = 2\operatorname{card} \Sigma'' + \operatorname{card} \Sigma'.$$

Finally, if $p(s) \in \Sigma''$, then $P_s = P_{p(s)}$, but s has a "twin"; if $p(s) \in \Sigma'$, then $P_s = 2P_{p(s)}$. By regrouping the equalities one has the desired formula. $\square$

N.B. In the computations, one must not forget that $P_s \geq 3$.

5.2 POINCARÉ RECURRENCE AND THE STABILITY LEMMA

The goal of this section is to prove two essential facts in the theory of foliations, namely, the Poincaré recurrence theorem and the stability lemma. Both are deduced from the existence of a good atlas, which is where we start.

Good atlas. Let M be a compact surface with a measured foliation. There exists a constant ϵ_0 and two finite covers $\{U_j\}_{j \in J}$, $\{V_j\}_{j \in J}$, by domains of charts, satisfying:

1. $M = \bigcup_{j \in J} (\operatorname{int} U_j)$.

2. For each $j \in J$, U_j is contained in V_j, and the faces of U_j are contained in the faces of V_j (see Figure 5.3).

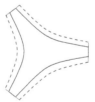

Figure 5.3.

3. Every point of the sides of U_j is a distance greater than ϵ_0 from the sides of V_j (all distances are measured along trajectories in the sense of the invariant measure μ).

4. Each singular point belongs to only one chart U_j.

5. The intersection of two charts U_{j_1} and U_{j_2} (resp. V_{j_1} and V_{j_2}) is a rectangle:

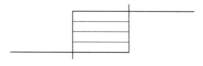

To satisfy the last condition, we choose a line field transverse to the foliation on the complement of the singularities and we insist that the charts are small enough that their faces are tangent to this line field.

THEOREM 5.2 (Poincaré recurrence). *Let M be a compact surface equipped with a measured foliation $(\mathcal{F}, \mu)$. Let α be an embedded arc of ∂M that is transverse to $\mathcal{F}$ at all points of its interior, and let x be one of its endpoints. Then the leaf L_x leaving from x either goes to a singular point or goes to the boundary.*

Proof. We will use the good atlas from above. We suppose that L_x does not reach a singularity. We truncate α so that $\mu(\alpha) = \epsilon < \epsilon_0$ and so, for every $y \in \alpha$, the leaf L_y does not end in a singularity. We claim that if L_x does not meet the boundary again, then we have an injective immersion $\Phi \colon \alpha \times \mathbb{R}_+ \to M$, where $\Phi(\{y\} \times \mathbb{R}_+) = L_y$ for each $y \in \alpha$.

Indeed, if P is a plaque of L_x in U_i, it is in the boundary of a strip of V_i. The strip has width ϵ and, by hypothesis on α, does not contain any singularities. If two plaques of L_x overlap, the strips in question glue together by the properties of a good atlas. Hence ϕ is an immersion. It is injective because $\Phi^{-1}(\alpha) = \alpha \times \{0\}$ and because each point of the image of Φ has only one leaf passing through it.

Let z be a *point of recurrence* of the leaf L_x, that is, a point z of L_x that, in some chart, is an accumulation point of the plaques of L_x not containing z. If $z \in U_i$ there are infinitely many strips of size ϵ that are components of $\operatorname{Im} \Phi \cap V_i$. But two distinct strips are disjoint—impossible.   □

COROLLARY 5.3. *If a leaf L of $\mathcal{F}$ is not closed in $M - \operatorname{Sing}(\mathcal{F})$, and if α is an arc transverse to $\mathcal{F}$ intersecting L, then $\alpha \cap L$ is infinite.*

In general, a leaf, or a collection of leaves, that is closed in the complement of the singularities is called an *invariant set*. It follows from Poincaré recurrence that each leaf of an invariant set (which is not the whole surface) is a *closed leaf*, that is, a closed loop or a leaf connecting a singularity to itself, or a *saddle connection*, that is, a leaf connecting two singularities.

Proof. It suffices to show that $\alpha \cap L$ cannot be a single endpoint of α; so assume for contradiction that $\alpha \cap L$ is one endpoint. We cut M along the interior of α to obtain a surface M' equipped with the induced foliation $\mathcal{F}'$. If C is the curve of $\partial M'$ arising from α, then near C the foliation $\mathcal{F}'$ gives the configuration of Figure 5.4. There are two singularities s_1 and s_2 corresponding to the endpoints of α, and there are two leaves L_1 and L_2 that come from L and that emanate from s_1.

By Poincaré recurrence, L_1 (resp. L_2) reaches a singularity of $\mathcal{F}'$ or the boundary of M'. If this boundary is C, by the hypotheses on α, we conclude that $L_1 = L_2$, which implies that L is closed—contradiction. Otherwise, considering M' contained in M, L_1 and L_2 reach singularities of $\mathcal{F}$ or the boundary of M; thus L is closed (contradiction).   □

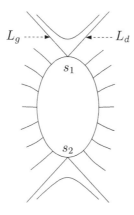

Figure 5.4.

The holonomy map. Let γ be a compact arc in a leaf, and let α, β be two disjoint transverse arcs each leaving from an endpoint of γ, both on the same side. Denote by L_t the leaf passing through $\alpha(t)$; $\alpha(0)$ and $\beta(0)$ are the endpoints of γ in L_0. We choose the parametrization in such a way that

$$\mu([\alpha(0), \alpha(t)]) = \mu([\beta(0), \beta(t)]) = t.$$

There is a *holonomy map*

$$h_\gamma \colon (\alpha, \alpha(0)) \to (\beta, \beta(0))$$

characterized by the following property: h_γ is continuous and if $h_\gamma(\alpha(t))$ is defined, we have $h_\gamma(\alpha(t)) \subset L_t$. In other words, the map is defined by following the leaves of the foliation from α to β. The invariance of the measure μ implies that h_γ is an isometry, that is to say, $h_\gamma(\alpha(t)) = \beta(t)$; we denote by $\{\gamma_t\}$ the continuous family of arcs such that $\gamma_0 = \gamma$, $\gamma_t \subset L_t$, and γ_t joins $\alpha(t)$ to $\beta(t)$.

LEMMA 5.4 (Stability lemma). *If h_γ is defined on the half-open interval $[\alpha(0), \alpha(t_0))$, then the points $\alpha(t_0)$ and $\beta(t_0)$ can be joined by an arc γ_{t_0} that is contained in a union of a finite number of leaves and singular points and that is the limit of the arcs γ_t, where $t \in [0, t_0)$.*

Furthermore, there exists an immersion $H \colon [0,1] \times [0, t_0] \to M$ that is C^∞ on the interior and such that $H([0,1] \times \{t\}) = \gamma_t$ for all $t \in [0, t_0]$.

The only obstructions to prolonging h_γ beyond $\alpha(t_0)$ are:

- $\alpha(t_0)$ *(resp. $\beta(t_0)$) is an endpoint of α (resp. β).*

- $\gamma(t_0)$ *contains a singularity.*

Proof. We again make use of the good atlas, and the notations U_j, V_j, ϵ_0. We may clearly reduce to the case where $t_0 < \epsilon_0$, where the arc $[\alpha(0), \alpha(t_0)]$ is contained in a chart V_{j_0}, and where the arc $[\beta(0), \beta(t_0)]$ is contained in a chart V_{j_1}. We

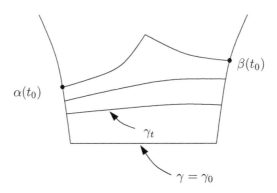

Figure 5.5.

then cover γ_0 by charts $U_{j_0} = U_0, U_1, \ldots, U_n = U_{j_1}$ (charts may be repeated). The numbering is chosen in such a way that for each i there is a plaque P_i^0 of U_i that is contained in $U_i \cap \gamma_0$ and that satisfies $P_i^0 \cap P_j^0 = \emptyset$, except when $|j - i| = 1$.

Consider the union $X_0 = \bigcup \{ P_0^t \mid t \in [0, t_0] \}$ of plaques of V_0 that intersect $[\alpha(0), \alpha(t_0)]$. A potential singularity of X_0 can only be found on the plaque $P_0^{t_0}$, for otherwise the holonomy map would not be defined on $[\alpha(0), \alpha(t_0))$. If we pass to the chart V_1, we find an intersection $X_0 \cap V_1$ that is a rectangle of width t_0, by the properties of a good atlas. We construct the union X_1 of plaques of V_1 that meet $X_0 \cap V_1$ and we continue in this way for the rest. $\square$

Remark 1. The stability lemma requires the invariant measure, in particular the existence of a good atlas. Figure 5.6 is a counterexample in the case where the measure has nontrivial holonomy.

Remark 2. The stability lemma remains true if γ_0 passes through singularities whose separatrices are on the side of γ_0 opposite from α and β.

COROLLARY 5.5. *We suppose that M is not the torus T^2. Let γ be a cycle of leaves (i.e., a simple closed curve that is a union of leaves and singularities). Either γ passes through singularities and there exist separatrices on both sides of γ, or γ belongs to a "maximal annulus" A whose interior leaves are cycles. In the latter case, any component of ∂A that is not in ∂M is a singular cycle.*

5.3   MEASURED FOLIATIONS AND SIMPLE CLOSED CURVES

Let M be a compact surface. We say that two measured foliations on M are *Whitehead equivalent* if they differ by

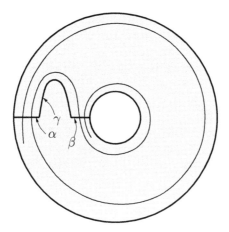

Figure 5.6. A counterexample to the stability lemma

- isotopy, and

- Whitehead operations:

(1)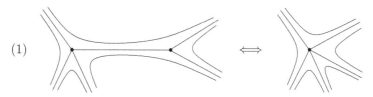

We will write $\mathcal{MF}(M)$—or simply $\mathcal{MF}$ when there is no ambiguity—to denote the set of Whitehead equivalence classes of measured foliations with permissible singularities. (We specify that if the two singularities are on the boundary, we only contract if the connecting leaf is contained in the boundary.)

(2)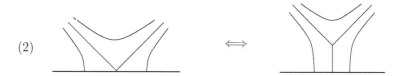

Recall that $\mathcal{S}$ denotes the set of homotopy classes (equivalently, isotopy classes) of simple closed curves that are piecewise C^∞, not homotopic to a point, and not isotopic to a curve of the boundary.

The map $I_*\colon \mathcal{MF} \to \mathbb{R}_+^{\mathcal{S}}$. Let $(\mathcal{F}, \mu)$ be a measured foliation and γ a closed curve. We set $\mu(\gamma) = \sup(\sum \mu(\alpha_i))$ where $\alpha_1, \ldots, \alpha_k$ are arcs of γ, mutually disjoint and transverse to $\mathcal{F}$, and where the sup is taken over all sums of this type. In other words, $\mu(\gamma)$ is the total variation of the y-coordinate along γ in an atlas that defines the measured foliation. This quantity is also denoted by Thurston as $\int_\gamma \mathcal{F}$. Let σ be an element of $\mathcal{S}$. We set

$$I(\mathcal{F},\mu;\sigma) = \inf_{\gamma \in \sigma} \mu(\gamma).$$

This is clearly an isotopy invariant. Moreover, if $(\mathcal{F},\mu)$ and $(\mathcal{F}',\mu')$ differ from each other by a Whitehead operation, then, for each curve $\gamma \in \sigma$ and each $\epsilon > 0$, there exists $\gamma' \in \sigma$ such that $|\mu(\gamma) - \mu'(\gamma')| < \epsilon$ (see Figure 5.7).

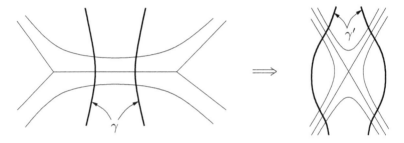

Figure 5.7.

Thus the above formula defines a function

$$I_* : \mathcal{MF} \to \mathbb{R}_+^{\mathcal{S}},$$
$$\langle I_*(\mathcal{F},\mu), \sigma \rangle = I(\mathcal{F},\mu;\sigma).$$

Quasitransverse curves. We now wish to find best representatives for homotopy classes of curves with respect to the map $I_* : \mathcal{MF} \to \mathbb{R}_+^{\mathcal{S}}$. This is analogous to finding geodesics with respect to a hyperbolic metric or minimal position representatives with respect to the geometric intersection number.

We say that a curve γ is *quasitransverse* to a foliation $\mathcal{F}$ if each connected component of $\gamma - \operatorname{Sing} \mathcal{F}$ either is a leaf or is transverse to $\mathcal{F}$. Further, in a neighborhood of a singularity, we insist that no transverse arc lies in a sector adjacent to an arc contained in a leaf, and that consecutive transverse arcs lie in distinct sectors. See Figure 5.8.

PROPOSITION 5.6. *Given any measured foliation $\mathcal{F}$, and any quasitransverse arc β, there does not exist a disk D with $\partial D = \alpha \cup \beta$, where α is an arc contained in a leaf of $\mathcal{F}$.*

Proof. Suppose that such a disk exists. Let $N \cong D^2$, the double of D along β. The disk N is endowed with a foliation with permissible singularities. But $\chi(N) > 0$, which contradicts the Euler–Poincaré formula.   □

Figure 5.8. A quasitransverse curve

Remark. In the case that α is a single point, we see that an immersed closed curve that is quasitransverse to $\mathcal{F}$ cannot be homotopic to a point.

PROPOSITION 5.7. *If γ is quasitransverse to $\mathcal{F}$, then*

$$\mu(\gamma) = I(\mathcal{F}, \mu; \sigma)$$

where σ is the homotopy class of γ.

Proof. Let $\gamma' \in \sigma$. If γ and γ' are disjoint, then γ and γ' bound an annulus A. By Poincaré recurrence, almost every leaf entering A at a point of γ meets the boundary again; by Proposition 5.6, it cannot meet γ again. Hence $\mu(\gamma) \leq \mu(\gamma')$.

If γ and γ' intersect, we proceed as follows. We begin by putting γ' in general position with respect to γ, in the sense that $\gamma' - \gamma$ is a finite number of open intervals; this is done by an approximation that changes the measure by an arbitrarily small amount. Since γ and γ' are homotopic, there exists an arc α' in γ' and an arc α in γ such that $\text{int } \alpha \cap \text{int } \alpha' = \emptyset$, and $\alpha \cup \alpha'$ bounds a disk D (Proposition 3.10). Almost every leaf entering D at a point of α meets the boundary again. Thus $\mu(\alpha) \leq \mu(\alpha')$. If $\gamma' = \alpha' \cup \beta'$, we may form $\gamma'' = \alpha'' \cup \beta''$, with $\alpha'' = \alpha$ and $\beta'' = \beta'$. We have $\mu(\gamma'') \leq \mu(\gamma')$ and $\pi_0(\gamma'' - \gamma) < \pi_0(\gamma' - \gamma)$. Thus, by induction on $\pi_0(\gamma' - \gamma)$, it follows that $\mu(\gamma') \geq \mu(\gamma)$. $\square$

To determine which elements of $\mathcal{S}$ contain quasitransverse curves, we require the following lemma about the holonomy map.

LEMMA 5.8 (Minimality criterion). *Let γ be a simple closed (connected) curve in the surface M and $(\mathcal{F}, \mu)$ a measured foliation. The following two assertions are equivalent:*

1. $\mu(\gamma) > I(\mathcal{F}, \mu; [\gamma])$.

2. *There exist two points x_0 and x_1 of γ that belong to the same leaf L and that satisfy*

$$\begin{aligned}
x_0 \cup x_1 &= \partial c, \text{ where } c \text{ is an arc of } L, \\
&= \partial c', \text{ where } c' \text{ is an arc of } \gamma, \text{ and} \\
c \cup c' &= \partial D, \text{ where } D \text{ is a 2-disk.}
\end{aligned}$$

Proof. By Proposition 5.7, to prove the lemma, one only needs to show that if there is no disk D as in the second statement of the lemma, then γ is isotopic to a quasitransverse curve of the same length.

We may suppose that $\gamma = \alpha_1 * \beta_1 * \cdots * \alpha_n * \beta_n$, where the arcs α_i are transverse to $\mathcal{F}$ and where the arcs β_j each lie in a finite union of leaves and singular points. We take the labeling to be cyclic, and we allow the β_i to be points. If we do not begin with such a decomposition of γ, we either obtain one in each chart by an isometric isotopy, or there exists a chart in which the second conclusion of the minimality criterion is visible and a length reducing modification leads to a finite decomposition.

Now, each β_k is in one of the configurations shown in Figure 5.9. In configuration (1), we can apply the stability lemma. In configurations (2) and (3), the arc β_k contains at least one singularity, and the stability lemma is not applicable. In configuration (4), the arc β_k does not contain any singularities.

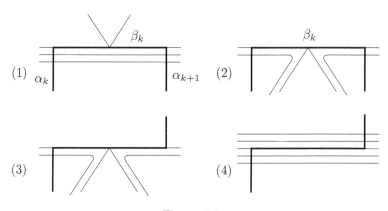

Figure 5.9.

In configuration (1), we see a disk as in the second conclusion of the minimality criterion. We claim that if, for all k, the arc β_k is not in configuration (1), then γ is isotopic to a quasitransverse curve of the same length; that is, $\mu(\gamma)$ is minimal by Proposition 5.7, and so the claim proves the minimality criterion. To prove the claim, we replace each configuration of type (4) by a transversal; each configuration of type (2) or (3) is modified as in Figure 5.10.   □

In the next proposition, a *spine* for a surface M is a 1-complex in M onto which the surface deformation retracts.

PROPOSITION 5.9. *Let γ be a simple closed (connected) curve in the surface M and $(\mathcal{F}, \mu)$ a measured foliation.*

1. *If $I(\mathcal{F}, \mu; [\gamma]) \neq 0$, there exists $(\mathcal{F}', \mu')$ equivalent to $(\mathcal{F}, \mu)$, such that γ is transverse to $\mathcal{F}'$ and avoids the singularities.*

2. *If $I(\mathcal{F}, \mu; [\gamma]) = 0$, there exists a foliation $(\mathcal{F}', \mu')$ that is equivalent to $(\mathcal{F}, \mu)$ and that satisfies one of the following two (nonexclusive) conditions:*

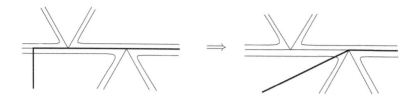

Figure 5.10.

a) γ is a cycle of leaves of $\mathcal{F}'$.

b) γ separates M into two components, say $M = M_1 \cup_\gamma M_2$, and for some $i \in \{1,2\}$ there is a spine Σ_i for M_i that is an invariant set of $\mathcal{F}'$.

Conclusion 2 can occur only if the set of connections between the singularities has cycles.

Remark 1. If we do not allow modification of $\mathcal{F}$, we obtain only the much weaker result that γ is homotopic to an immersion that is quasitransverse to $\mathcal{F}$ and that is a limit of embeddings.

Remark 2. Figure 5.11 illustrates the situation of case *2(b)* of the proposition. The foliation of the surface of genus two is obtained by "enlarging" the curve C (see Section 5.4).

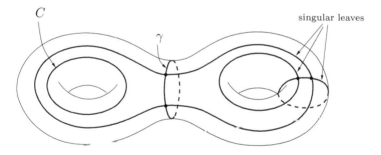

Figure 5.11. Proposition 5.9 case *2(b)*

Proof of Proposition 5.9. As in the proof of the minimality criterion (Lemma 5.8), we write γ as $\alpha_1 * \beta_1 * \cdots * \alpha_n * \beta_n$. As in the same proof, the arcs β_k of type (4) are replaced by transverse arcs, and those of types (2) and (3) may be supposed to have singularities at both endpoints.

At this point, either γ is a cycle of leaves, which gives Conclusion *2(a)* of the proposition, or it is possible to shrink each full leaf contained in γ to a point (first arrow of Figure 5.12). By then blowing up the resulting singular points as shown in the second arrow of Figure 5.12, we eliminate arcs of types (2) and (3), thus reducing to the situation where each of the arcs β_k is of type (1). From there, the induction is done on the number of arcs of γ contained in a leaf. If there are none, γ is transverse to the foliation, which is conclusion 1 of the proposition.

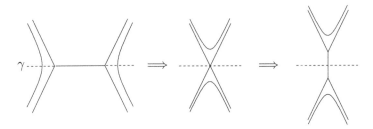

Figure 5.12.

Otherwise, consider β_1, which is in configuration (1) by assumption. Applying the stability lemma to the arcs β_1, α_0, and α_1, we obtain an immersion h of a rectangle R. The induced foliation $\widehat{\mathcal{F}} = h^{-1}(\mathcal{F})$ has all of its singularities in the same arc λ of the boundary. We denote by $\widehat{\beta}_1, \ldots, \widehat{\beta}_m$ the arcs of $\widehat{\gamma} = h^{-1}(\gamma)$ that are in the leaves of $\widehat{\mathcal{F}}$ (horizontal arcs). Let us say that $\widehat{\beta}_1$ is an arc closest to the singularities in the sense of the transverse measure. It follows that the component of $\widehat{\gamma}$ that contains $\widehat{\beta}_1$ bounds a subrectangle R' that is minimal. Also, we see that $h|_{\text{int } R'}$ is an embedding disjoint from γ.

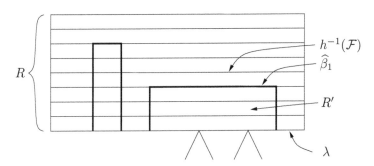

Figure 5.13.

If R' does not contain any singularities of $\widehat{\mathcal{F}}$, a neighborhood of $h(R')$ is the support for an isotopy of γ that gets rid of $\widehat{\beta}_1$. Even if $h(\widehat{\beta}_1) = \beta_1$, the application of this isotopy leads to a situation where, in the new rectangle R associated with β_1, the new $\widehat{\gamma}$ has fewer horizontal arcs.

If R' has a singularity, then, because of the transverse measure, it is easy to see that $h(\widehat{\beta}_1)$ is an arc β_k distinct from β_1 (otherwise the width of R' would be the same as that of R).

By the above reasoning we may suppose that, up to cyclically relabeling the arcs, we have:

1. $h|_{R-\lambda}$ is an embedding,

2. $\left(h|_{\text{int}(R)}\right) \cap \gamma$ is empty, and

3. $h(\lambda) \cap \beta_k$ is empty for all k.

We first brush aside the following simple cases *(A)* and *(B)*, where there are visible isotopies and Whitehead operations that reduce the number of arcs of γ contained in a leaf.

(A) λ does not contain a singularity. See Figure 5.14.

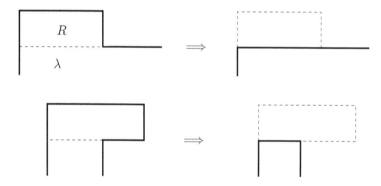

Figure 5.14.

(B) λ contains singularities and R is embedded. The isotopy across R replaces β_1 with an arc of type 2. We then perform the procedure from the beginning of the proof.

We may now assume that R is not embedded, and so $h(\lambda)$ has double points. Viewed as a singular chain, λ is written as a composition:

$$\lambda = \mu_0 * \lambda_1 * \cdots * \lambda_q * \mu_1$$

where μ_0 (resp. μ_1) is an arc of a leaf joining a point of α_1 (resp. α_2) to a singularity and where λ_i ($1 \le i \le q$) is an arc of a leaf joining two singularities. Some of these arcs may be a single point and several may belong to the same leaf. In any case, λ has in R an approximation that is an embedded arc only meeting α_1 and α_2 at their endpoints. Because of this, each leaf carries at most two arcs

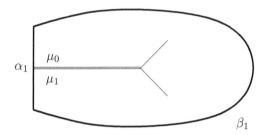

Figure 5.15.

of λ. In particular, neither μ_0 nor μ_1 may belong to the same leaf as a λ_j. If $\mu_0 \cap \mu_1$ is not a single one of their endpoints, then $\alpha_1 = \alpha_2$ (i.e., $\gamma = \alpha_1 * \beta_1$) and we have the configuration of Figure 5.15.

We will say that λ_j is *simple* if λ_j does not cover the same leaf as some other $\lambda_{j'}$. We say that μ_0 and μ_1 are simple if one does not have the configuration of Figure 5.15 (Note that μ_0 and μ_1 are both simple or both not simple).

Denote by Λ the one-dimensional complex

$$\bigcup_{i=1}^{q} \lambda_i;$$

this is an invariant set of the foliation $\mathcal{F}$. If M is closed, each Whitehead operation of Λ lifts to a Whitehead operation of $\mathcal{F}$ (the terminology for foliations was chosen because of this remark).

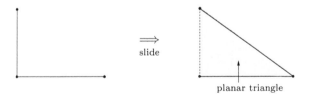

Claim. Assume M is closed. If one of the arcs λ_i, μ_0, or μ_1 is simple, then there exists a foliation $\mathcal{F}'$ that is equivalent to $\mathcal{F}$ and that is equal to $\mathcal{F}$ in the complement of a neighborhood of Λ, and for which the limit arc λ' of the domain of deformation of R' of β_1 has fewer double simplices (edges or vertices).

To prove the claim, we slide the simple arc on its predecessor or on its successor. Figure 5.16 exhibits this operation when μ_0 is simple.

If the claim is applicable, we reduce by induction to case *(B)*; otherwise, we find ourselves in the following situation.

(C) All of the arcs λ_i, μ_0, and μ_1 are doubled.

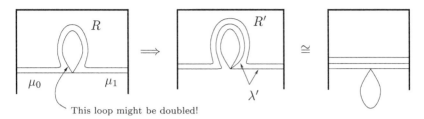

Figure 5.16.

In this case, the closure of R in the surface is a regular neighborhood of the complex Λ and γ is its boundary; we thus have conclusion *2(b)* of the proposition.

In the case where M is closed, the proof of the proposition is completed by induction on the number of segments of the decomposition of γ. The case of surfaces with boundary is analogous, but one must pay attention to the Whitehead operations permitted. $\square$

Remark. The preceding proposition does not admit a reasonable generalization to the case of a system with k embedded curves $\gamma_1, \ldots, \gamma_k$, except if $I(\mathcal{F}, \mu; [\gamma_1]) \neq 0, \ldots, I(\mathcal{F}, \mu; [\gamma_{k-1}]) \neq 0$, and $I(\mathcal{F}, \mu; [\gamma_k])$ is possibly zero.

5.4 CURVES AS MEASURED FOLIATIONS

We start by explaining a procedure for going from a measured foliation on a subsurface to a measured foliation on the whole surface. Then we specialize to the case where the subsurface is an annular neighborhood of a simple closed curve.

The enlarging procedure. Let M_0 be a submanifold of dimension 2 in M such that $M - M_0$ does not have any contractible components. Let Σ be a spine of $\overline{M - M_0}$. By hypothesis, none of the components of Σ are contractible. Thus, perhaps after collapsing the 1-simplices that have a free vertex, each singularity of Σ has at least three branches leaving from it.

We may construct a surjective map $j: M_0 \to M$ such that

- j is a (piecewise differentiable) immersion.

- $j|_{\text{int } M_0}$ is a diffeomorphism onto $M - \Sigma$.

- $j(\partial M_0 - \partial M) = \Sigma$.

- j is the identity outside of a small collar neighborhood of $\partial M_0 - \partial M$.

Let $\mathcal{F}_0$ be a measured foliation on M_0 such that each component of $\partial M_0 - \partial M$ is an invariant set of $\mathcal{F}_0$. We may then define $\mathcal{F} = j_*(\mathcal{F}_0)$, which is a measured foliation on M satisfying

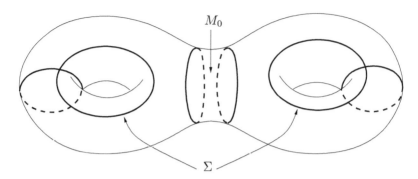

Figure 5.17.

- Σ is an invariant set of $\mathcal{F}$.

- $j|_{\text{int}(M_0)}$ conjugates $\mathcal{F}_0|_{\text{int}(M_0)}$ and $\mathcal{F}|_{(M-\Sigma)}$ as measured foliations.

We say that $\mathcal{F}$ is obtained from $\mathcal{F}_0$ by *enlarging* M_0.

We remark that if Σ' is another spine of $\overline{M - M_0}$, then Σ' is obtained from Σ by Whitehead operations and isotopies (see Appendix B). We conclude that the class of $\mathcal{F}$ depends only on that of $\mathcal{F}_0$. We have therefore defined a map

$$\mathcal{MF}(M_0, \partial M_0 - \partial M) \to \mathcal{MF}(M)$$

for which the domain is the subset of $\mathcal{MF}(M_0)$ consisting of the foliations where every component of $\partial M_0 - \partial M$ is an invariant set.

LEMMA 5.10. *Let μ_0 and μ be transverse invariant measures for $\mathcal{F}_0$ and $\mathcal{F}$. Let γ be a simple curve in M. Then $I(\mathcal{F}, \mu; [\gamma]) = \inf \mu_0(\gamma' \cap M_0)$, where γ' is isotopic to γ.*

Proof. The lemma is a consequence of the following remark: for each curve C, there exists a curve C', isotopic to C, such that $C' \cap M_0 = j^{-1}(C)$. $\qquad\square$

The inclusion $\mathbb{R}_+ \times \mathcal{S} \hookrightarrow \mathcal{MF}$. Let $C \in \mathcal{S}$, $\lambda \in \mathbb{R}_+^*$. Consider a tubular neighborhood M_0 of C, which we foliate by circles parallel to C. We equip the foliation with an invariant transverse measure μ_0 such that the width of the annulus M_0 is λ. This measured foliation of M_0 is unique up to isotopy. We denote by $F_{\lambda,C}$ a foliation obtained from the latter by enlarging and by μ its transverse measure.

PROPOSITION 5.11. *Let γ be a simple curve in M. Then we have*

$$I(\mathcal{F}_{\lambda,C}, \mu; [\gamma]) = \lambda\, i(C, \gamma).$$

Proof. Let α be a component of $\gamma \cap M_0$. If α goes from one boundary of M_0 to the other, then $\mu_0(\alpha) \geq \lambda$. We deform α by isotopy until it is transverse to the foliation; then $\alpha \cap C$ is one point and $\mu_0(\alpha) = \lambda$. If α intersects only one component of the boundary, then γ is isotopic to a curve γ' whose intersection with M_0 has one fewer component. Applying Lemma 5.10, we have the inequality

$$I(\mathcal{F}_{\lambda,C}, \mu; [\gamma]) \geq \lambda \, i(C, \gamma).$$

The equality is obtained by considering the case where γ has minimal intersection with C, for then we have

$$\mu_0(\gamma) = \lambda \, i(C, \gamma).$$

$\square$

The preceding proposition implies that the following diagram is commutative:

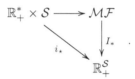

As i_* is injective (by Proposition 3.17), $\mathbb{R}_+^* \times \mathcal{S} \to \mathcal{MF}$ is also an injection.

Appendix B

Spines of Surfaces

by Valentin Poénaru

Let N be a compact, connected manifold of dimension 2, with a nonempty boundary. If N is triangulated and if $L_1 \subset L_2 \subset N$ are two subcomplexes, we say that we pass from L_1 to L_2 by a *dilation* of dimension n if there exists an n-simplex τ of N and a face τ' of τ such that

$$L_2 - L_1 = \operatorname{int} \tau \cup \operatorname{int} \tau'$$

(here int designates the open cell). The inverse operation is called *collapsing*. If one passes from L' to L'' by a sequence of dilations, then one can do so in an ordered way, such that the sequence of respective dimensions is nondecreasing.

A *slide* is a sequence of collapses and dilations

$$L'' = C_n D_n C_{n-1} D_{n-1} \cdots C_1 D_1(L') \tag{B.1}$$

where $\dim(L'') = \dim(L') = 1$, $\dim(C_i) = \dim(D_i) = 2$, and $\operatorname{supp}(C_i) = \operatorname{supp}(D_i)$. More generally, if $L', L'' \subset N$ are two complexes of dimension 1, we say that $L' \to L''$ is a *slide* if there exists a triangulation of N in which (B.1) is realized.

A subpolyhedron $L \subset N$ is a *spine* if, for a particular triangulation, N collapses to L.

THEOREM B.1. *Let Σ_1 and Σ_2 be two 1-complexes of N having no free ends. If Σ_1 and Σ_2 are two spines of N, we may pass from Σ_1 to Σ_2 by a sequence of slides and isotopies.*

The theorem is a consequence of the following lemmas.

LEMMA B.2. *Isotopies and slides transform a spine into a spine.*

LEMMA B.3. *Let Σ be a spine of N and L a simple arc of N that meets Σ only at its endpoints. There exists a continuous map $\varphi \colon D^2 \to N$ and a decomposition of ∂D^2 into two segments: $\partial D^2 = A \cup B$, $\partial A = \partial B$, $\operatorname{int} A \cap \operatorname{int} B = \emptyset$, such that:*

1. $\varphi|_A$ is a homeomorphism onto L.

2. $\varphi|_{\operatorname{int} D^2}$ is a smooth embedding into $N \setminus \Sigma$.

3. $\varphi(B) \subset \Sigma$.

The proofs of these two lemmas are left as an exercise.

LEMMA B.4. *For a triangulation of N, consider sequences of sub-complexes*

$$X^0 \subset X^1 \subset \cdots \subset X^n$$

and

$$Y^0 \subset Y^1 \subset \cdots \subset Y^n$$

having the following properties:

1. X^0 *and* Y^0 *are spines of* N.

2. The transformations $X^{i-1} \subset X^i$, $Y^{i-1} \subset Y^i$ *are dilations of dimension 2.*

3. X^n *is the same subcomplex of* N *as* Y^n.

Then there exists a sequence of subcomplexes $Z^0 \subset Z^1 \subset \cdots \subset Z^{n-1}$ *such that:*

4. Z^0 *is obtained from* Y^0 *by a slide (in particular, Z^0 is a spine).*

5. $Z^{n-1} = X^{n-1}$.

6. The transformations $Z^{i-1} \subset Z^i$ *are dilations of dimension 2.*

Proof. Let σ be a 2-simplex of N that corresponds to the dilation $X^{n-1} \subset X^n$, and let σ_1, σ_2, and σ_3 be its three faces. Denote by P_i the vertex opposite σ_i. Suppose also that σ_1 is the free face of the collapse $X^n \searrow X^{n-1}$ and that σ_j is the free face of the collapse $Y^{i_0} \searrow Y^{i_0-1}$:

$$Y^{i_0} - Y^{i_0-1} = \operatorname{int} \sigma \cup \operatorname{int} \sigma_j.$$

If $j = 1$, the lemma follows immediately; thus suppose that $j = 2$.

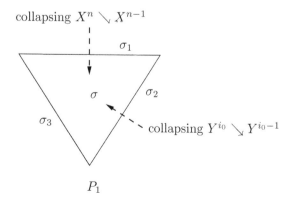

$$P_1$$

Since σ_1 is a free face in $Y^n = X^n$, this edge is not in the boundary of another 2-simplex of Y^{i_0-1}. Thus $\sigma_1 \subset Y^0$. Similarly, by property 2, the 0-skeleton of $X^n = Y^n$ is contained in X^0 and in Y^0. Therefore

$$\partial\sigma \cap Y^0 = \left\{ \begin{array}{l} \sigma_1 \cup \sigma_3, \text{ or} \\ \sigma_1 \cup P_1. \end{array} \right.$$

Let $Z^0 = (Y^0 - \text{int}\,\sigma_1) \cup \sigma_2$. If $\partial\sigma \cap Y^0 = \sigma_1 \cup \sigma_3$, it is evident that one can pass from Y^0 to Z^0 by a slide. If $\partial\sigma \cap Y^0 = \sigma_1 \cup P_1$, we may apply Lemma B.3 with $\Sigma = Y^0$ and $L = \sigma_3$. This therefore permits us to conclude, in this case, that the transformation $Y^0 \to Z^0$ is a slide. Thus point 4 is verified. The constructions of $Z^1 \subset \cdots \subset Z^{n-1}$ to ensure points 5 and 6 are left to the reader.   □

LEMMA B.5. *Let L_1, L_2 be two complexes of dimension 1 in N, with no free ends. Let L'_1, L'_2 be complexes obtained by dilations of dimension 1 from L_1 and L_2, respectively. If it is possible to pass from L'_1 to L'_2 by slides and isotopies, then the same is true for L_1 and L_2.*

This is an easy exercise. From these four lemmas, we can deduce the theorem without difficulty.

Exposé Six

The Classification of Measured Foliations

by Albert Fathi

The goal of this exposé is to classify measured foliations on a closed, orientable surface. What is more, we give coordinates for the space of measured foliations that are analogous to Fenchel-Nielsen coordinates for the Teichmüller space of a surface (see [FM11]). As in the case of curves, we will reduce to the case of foliations on pairs of pants. To do this, we choose curves $K_1, \ldots, K_{3g-3}$ that decompose the surface into pairs of pants.

If $(\mathcal{F}, \mu)$ is a foliation such that $I(\mathcal{F}, \mu; [K_j]) \neq 0$ for all j, we can perform isotopies and Whitehead operations in order to reduce to the case where the K_j are transverse to $\mathcal{F}$. Such a foliation is classified by the measures of the curves K_j and by the twists about these curves, which themselves are expressed according to measures of certain curves (see Appendix C).

In the case where the lengths of some of the K_i are zero, we can hope to modify $\mathcal{F}$ so that these curves will all be cycles of leaves, and then to classify the foliations as above. This is, unfortunately, not always possible. Here is an example on a surface M of genus 3.

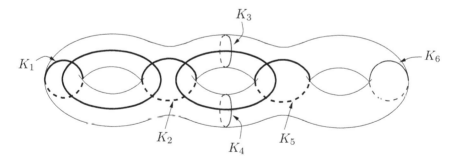

Figure 6.1.

In the example, the curves drawn in bold are singular leaves, and all other leaves are curves isotopic to K_6. This foliation is obtained by starting with a foliation on an annulus A around K_6 and collapsing $\overline{M - A}$ onto a spine. In this operation, the two pairs of pants that contain K_1 are collapsed onto the union of three closed curves. It is impossible to modify this foliation by Whitehead

operations and isotopies so that the curves K_3 and K_4 become cycles of leaves—
there are always points where these curves are tangent to the foliation. As a
consequence, the foliation does not restrict nicely to the pairs of pants.

We are thus obliged to take this kind of phenomenon into account. That is
why we have introduced the operation of enlargement (see the preceding exposé):
a foliation $\mathcal{F}$ of a surface M is obtained by enlargement of a foliation $\mathcal{F}_0$ of a
subsurface M_0 (with boundary) if $\mathcal{F}$ is the image of $\mathcal{F}_0$ under a map $M \to M$
obtained by extending a collapse of $\overline{M - M_0}$ onto a spine. Such a foliation is
essentially "carried" by M_0 since the transverse lengths of curves contained in
$M - M_0$ are zero.

Using this operation, we can find canonical ("normal") forms of foliations
and proceed to the classification.

6.1   FOLIATIONS OF THE ANNULUS

By the Euler–Poincaré formula, a measured foliation on $S^1 \times [0, 1]$ has no singularities. If $S^1 \times \{0\}$ is a leaf, then by the stability lemma all the leaves are closed
curves. If $S^1 \times \{0\}$ is transverse to the foliation, then all the leaves go from one
boundary to the other. Thus, if (θ, x) denotes the coordinates of $S^1 \times [0, 1]$, all
measured foliations of the annulus are isotopic to those associated to $\lambda\, d\theta$ or to
$\lambda\, dx$, where $\lambda \in \mathbb{R}^\star$.

We want to give a classification of measured foliations of the annulus A
modulo the action of the group $\mathrm{Diff}_0(A$ rel $\partial A)$ of diffeomorphisms that are
isotopic to the identity relative to the boundary. We choose once and for all an
arc γ joining the two boundary components and an arc $\overline{\gamma}$ differing from γ by a
twist in the positive direction (Figure 6.2).

If $(\mathcal{F}, \mu)$ is a measured foliation on A, we set

$$
\begin{aligned}
m &= \mu(S^1 \times \{0\}) = \mu(S^1 \times \{1\}) = I(\mathcal{F}, \mu; [S^1 \times \{0\}]), \\
s &= \inf\{\mu(\gamma') : \gamma' \text{ isotopic to } \gamma \text{ with endpoints fixed}\}, \\
t &= \inf\{\mu(\gamma') : \gamma' \text{ isotopic to } \overline{\gamma} \text{ with endpoints fixed}\}.
\end{aligned}
$$

LEMMA 6.1. *A triple (m, s, t) of three positive numbers is associated to a
measured foliation of A if and only if (m, s, t) belongs to $\partial(\nabla \le)$ (boundary of
the triangle inequality).*

Proof. We consider the hatched triangle T_1 in Figure 6.2. Having done an
isotopy such that γ and $\overline{\gamma}$ are transverse to the foliation, any leaf that enters
through one edge of the triangle must leave through one of the others. In this
situation, it is clear that the three measures of the edges form a triple belonging
to $\partial(\nabla \le)$. It is clear also that this condition is the only one that needs to be
satisfied so that a triple is associated to a measured foliation.   □

If $m = 0$, then each curve of the boundary is a leaf and, with the coordinates
(θ, x), the foliation is isotopic (rel ∂A) to $t\, dx = s\, dx$. This case being excluded,
the foliation is transverse to the boundary.

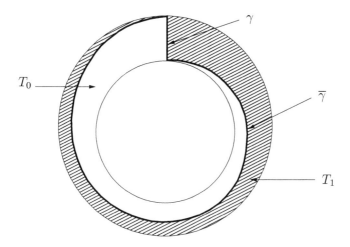

Figure 6.2.

PROPOSITION 6.2. (Classification of foliations of an annulus) *Let $\mathcal{F}$ and $\mathcal{F}'$ be two measured foliations on A that are transverse to ∂A and that coincide on ∂A (we mean equality of the induced measures; in particular, $m(\mathcal{F}) = m(\mathcal{F}')$). Then $\mathcal{F}$ and $\mathcal{F}'$ are isotopic by an isotopy that is constant on ∂A if and only if $(s,t)(\mathcal{F}) = (s,t)(\mathcal{F}')$.*

Proof. Only the sufficiency is nontrivial. We deform $\mathcal{F}$ and $\mathcal{F}'$ until

$$s = \mu(\gamma) = \mu'(\gamma),$$
$$t = \mu(\overline{\gamma}) = \mu'(\overline{\gamma}).$$

Then γ and $\overline{\gamma}$ are transverse to the two foliations, unless one of these arcs is a leaf. In the case of transversality, a second isotopy makes the measures induced on γ (resp. $\overline{\gamma}$) coincide. Then the foliations coincide on the boundary of each of the triangles T_0 and T_1. We know that such data on the boundary of a disk has a unique extension (for example, by Theorem 2.1). □

6.2 FOLIATIONS OF THE PAIR OF PANTS

We denote the pair of pants, or, disk with two holes, by P^2.

LEMMA 6.3. *For a foliation of P^2 with permissible singularities (Exposé 5) either there is only one singularity with four separatrices or there are two singularities, each with three separatrices.*

The lemma follows from the Euler–Poincaré formula.

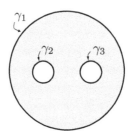

Figure 6.3. P^2 = pair of pants (disk with two holes)

Good foliations.   We say that $\mathcal{F}$ is a *good foliation* of P^2 if no component of ∂P^2 is a smooth leaf of $\mathcal{F}$ (a *smooth leaf* is one that does not contain any singularities).

LEMMA 6.4. *Let $\mathcal{F}$ be a measured foliation of P^2. Then*

1. *Every leaf is closed in the complement of the singularities. In other words, each leaf is an invariant set.*

2. *If, further, $\mathcal{F}$ is a good foliation, there are no cycles of leaves interior to P^2.*

Proof. 1. Let L be a leaf of $\mathcal{F}$ that is not closed in $P^2 - \mathrm{Sing}\,(\mathcal{F})$. It enjoys Poincaré recurrence and so one can find an arc β on L and a transversal α so that $\alpha \cup \beta$ is a simple closed curve. We have two possible configurations (Figure 6.4).

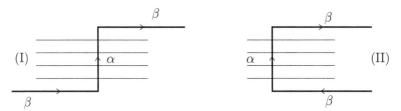

Figure 6.4.

By Corollary 5.3 (of Poincaré recurrence), L intersects α infinitely many times. Thus, configuration (II) reduces to configuration (I).

In configuration (I), we can approximate $\alpha \cup \beta$ by a closed curve γ that is transverse to $\mathcal{F}$ and that intersects L infinitely many times. By Proposition 5.6 the curve γ does not bound a disk. Therefore γ, together with some component γ_1 of ∂P^2, bounds an annulus. Since measured foliations of an annulus are "standard," each leaf that intersects γ also intersects γ_1. This implies that L cannot intersect γ infinitely many times, a contradiction.

2. If γ is an interior cycle, $\gamma \cup \gamma_1$ bounds an annulus A. In the neighborhood of γ in A the leaves are smooth and closed and, by the stability lemma, γ_1 is a smooth closed leaf, which is a contradiction. □

COROLLARY 6.5. *Every leaf of a good foliation of P^2 goes from the boundary to the boundary, from the boundary to a singularity, or from a singularity to a singularity.*

Reduced good foliations. A *reduced good foliation* of P^2 is a good foliation of P^2 satisfying the following conditions:

(i) If a component of the boundary is a transversal, it contains no singularities.
(ii) The singularities on the boundary are simple (three separatrices).
(iii) There are no connections between two singularities where at least one is interior.

Let $\mathcal{MF}_0(P^2)$ be the set of equivalence classes of good measured foliations of P^2.

LEMMA 6.6. *In each class of $\mathcal{MF}_0(P^2)$, there exists a unique reduced good foliation up to isotopy.*

Proof. We can secure property (i) immediately. Then, by part 2 of Lemma 6.4, if a foliation admits two simple singularities connected by two distinct arcs α_1 and α_2, then $\alpha_1 \cup \alpha_2$ is a component of the boundary. Thus, for an unreduced foliation, there is only one way to reduce (up to isotopy). □

PROPOSITION 6.7. (Classification of good foliations of a pair of pants) *The function $\mathcal{MF}_0(P^2) \to \mathbb{R}^3_+$, which to a good measured foliation $(\mathcal{F}, \mu)$ associates the triple*

$$(m_1, m_2, m_3) = (\mu(\gamma_1), \mu(\gamma_2), \mu(\gamma_3)),$$

induces a bijection of $\mathcal{MF}_0(P^2)$ onto $\mathbb{R}^3_+ - \{0\}$.

Proof. We begin by describing a right inverse. The construction depends on the position of the triple with respect to the triangle inequality; to each simplex, we associate one topological configuration. These are given in Figures 6.5 and 6.6 for the six types of simplices.

We remark that if we decompose these figures along the separatrices, we obtain foliated rectangles where the widths (i.e., the largest measures of transversals) are determined by the triple. For example in configuration (1), the widths of the three rectangles are

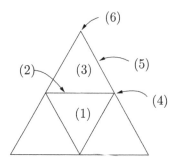

Figure 6.5.

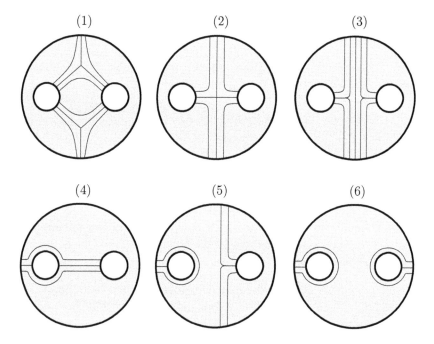

Figure 6.6.

$$a_{12} = \frac{1}{2}(m_1 + m_2 - m_3),$$

$$a_{13} = \frac{1}{2}(m_1 + m_3 - m_2),$$

$$a_{23} = \frac{1}{2}(m_2 + m_3 - m_1).$$

In configuration (3), we have the formulas

$$a_{11} = \frac{1}{2}(m_1 - (m_2 + m_3)),$$

$$a_{12} = m_2,$$

$$a_{13} = m_3.$$

It is easy to see that up to renumbering the boundary components, these figures represent all the possibilities up to isotopy for the separatrices of a reduced foliation; all the other configurations are ruled out by the Euler–Poincaré formula. We deduce right away that two foliations giving the same triple are isotopic. □

We consider now the case where some curves of the boundary are smooth leaves. We can immediately see two constructions of such foliations: adjoin a smoothly foliated annulus to the boundary of a good foliation (along a nonsmooth leaf), or enlarge one or more boundary components of the pair of pants. We now see that these constructions account for all foliations of a pair of pants.

PROPOSITION 6.8. (Classification of measured foliations of a pair of pants) *Let $\mathcal{F}$ be a measured foliation of P^2. Then $\mathcal{F}$ is obtained by enlargement of a foliation $\mathcal{F}_0$ of a submanifold P_0 where each connected component C of P_0 is*

(i) a pair of pants, in which case the foliation $\mathcal{F}_0|_C$ is a good foliation, or

(ii) a collar neighborhood of a curve of ∂P^2, in which case $\mathcal{F}_0|_C$ is a foliation by circles.

Proof. If no boundary component is smooth, we take $P_0 = P^2$. Otherwise, we consider a smooth leaf γ_1 in ∂P^2. We consider the maximal "annulus" A associated to γ_1 by the stability lemma. If $A = P^2$, we take P_0 to be a collar neighborhood of γ_1 foliated by circles, where the $\mathcal{F}_0$-width is the $\mathcal{F}$-width of A.

If $A \neq P^2$, there exists a leaf L of $\mathcal{F}$ in the interior of P^2 that belongs to the topological frontier of A. Its closure $\overline{L}$ contains at least one singularity.

If $\overline{L}$ contains one singularity s_0, then $\overline{L}$ is a cycle of leaves forming a Jordan curve which bounds a true annulus A' foliated by circles. The domain $\overline{P^2 - A'}$, which is a pair of pants—possibly pinched if s_0 belongs to ∂P^2—is foliated with fewer smooth leaves in its boundary. We kill any other smooth components in the same fashion.

If $\overline{L}$ connects distinct singularities s_0 and s_1, then the singularities are simple (Figure 6.7) and some other leaf L' leaves s_0 in the frontier of A. If L' returns to s_0, $\overline{L'}$ is an embedded cycle. Otherwise L' goes to s_1 and $\overline{L \cup L'}$ is a Jordan cycle. In either case, we continue as above. □

Propositions 6.7 and 6.8, taken together, constitute the classification of measured foliations of a pair of pants.

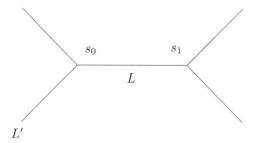

Figure 6.7.

6.3   THE PANTS SEAM

We will need a technical ingredient that will allow us to give coordinates for a measured foliation, using its image on each piece of a pair of pants decomposition of the given surface.

For each component C of the boundary of the pair of pants P^2, and for each type of good foliation on P^2, we choose a quasitransverse arc that has endpoints in C, and that is essential (not homotopic to an arc of the boundary). We call it the *pants seam*.[1] For $C = \gamma_1$ and for each type of good foliation, we choose the arc indicated in bold in Figures 6.8–6.13.

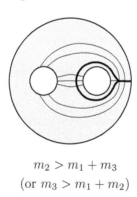

$$m_2 > m_1 + m_3$$
$$(\text{or } m_3 > m_1 + m_2)$$

Figure 6.8. Generic case

Length of the pants seam.   We remark that a pants seam realizes the minimum of the length of an essential arc going from γ_1 to γ_1 (by quasitransversality). Its

---

[1] Translators' note: In the original text, this arc was called the *arc jaune* ("yellow arc"), presumably a reference to the color of the chalk used to draw it on the chalkboard at the seminar.

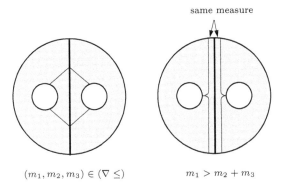

$(m_1, m_2, m_3) \in (\nabla \leq)$ $m_1 > m_2 + m_3$

Figure 6.9. Generic case

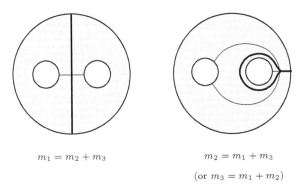

$m_1 = m_2 + m_3$ $m_2 = m_1 + m_3$

(or $m_3 = m_1 + m_2$)

Figure 6.10. Case where $(m_1, m_2, m_3) \in \partial(\nabla \leq)$ with $m_1 \neq 0, m_2 \neq 0, m_3 \neq 0$

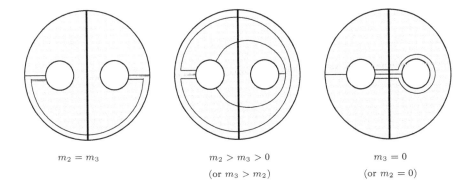

$m_2 = m_3$ $m_2 > m_3 > 0$ $m_3 = 0$

(or $m_3 > m_2$)   (or $m_2 = 0$)

Figure 6.11. Case where $m_1 = 0$

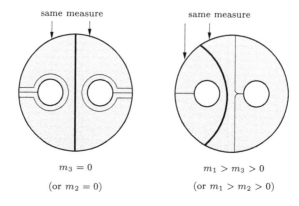

Figure 6.12. Case where $m_2 = 0$ (or $m_3 = 0$)

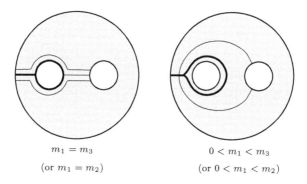

Figure 6.13. Case where $m_2 = 0$ (or $m_3 = 0$)

length (in the sense of the transverse measure) is given by the formula

$$l_1 = \sup\left(\tfrac{m_2+m_3-m_1}{2},0\right) +$$
$$\sup\left(\tfrac{m_2-m_1-m_3}{2},0\right) +$$
$$\sup\left(\tfrac{m_3-m_1-m_2}{2},0\right).$$

The proof is done by examining *all* the cases of the figure.

Definition of the arc A. The pants seam demarcates two arcs on γ_1, with one of them being possibly reduced to a point. We denote by A the one that is on the same side as γ_2 with respect to the pants seam (one should denote A by A_{12}, but there will not be any ambiguity later). Its length a is given by the formula

$$a = \sup\left(\frac{m_2 + m_1 - m_3}{2},0\right) - \sup\left(\frac{m_2 - m_1 - m_3}{2},0\right).$$

6.4   THE NORMAL FORM OF A FOLIATION

Let M be a closed surface of genus $g \geq 2$, and let $K_1, \ldots, K_{3g-3}$ be a family of curves that separates M into pairs of pants. We denote by $\{R_j\}$ the $2g - 2$ pairs of pants. Each R_j is the closure of one of the components of $M - \cup K_i$.

We will need the following technical condition: each R_j must be the image of an embedded pair of pants; in other words, no K_i lies in the same pair of pants on its two sides. With this condition, we say that we have a *permissible decomposition* of M.

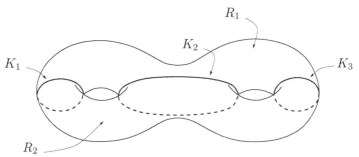

Figure 6.14.

For each $i = 1, \ldots, 3g-3$, we choose a tubular neighborhood $K_i \times [-1, 1] \subset M$ of K_i; if $i \neq j$, the neighborhoods K_i and K_j are disjoint. We denote by $\{R'_j\}_{1 \leq j \leq 2g-2}$ the connected components of $M - \cup(K_i \times (-1, 1))$.

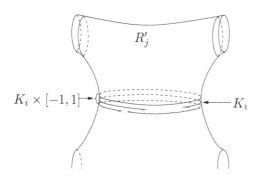

Figure 6.15.

Definition.   Let M_0 be a compact submanifold of M of dimension 2 and $\mathcal{F}_0$ a measured foliation of M_0. We say that $(M_0, \mathcal{F}_0)$ is in *normal form* (with respect to the data of the preceding paragraph) if the following conditions are satisfied.

(1) Each component of ∂M_0 is a cycle of leaves.

(2) $M_0 \cap R'_j$ is empty or equal to R'_j; in the latter case $\mathcal{F}_0|_{R'_j}$ is a good foliation.

(3) $M_0 \cap K_i \times (-1, 1)$ is equal to one of the following:

- The empty set.
- $K_i \times (-1, 1)$. In this case, $\mathcal{F}_0$ is transverse to the circles $K_i \times \{t\}$, $t \in [-1, 1]$. We also remark that in this case M_0 contains the pairs of pants adjacent to the annulus.
- $K_i \times [-\frac{1}{2}, \frac{1}{2}]$. In this case the foliation has the $K_i \times \{t\}$, $t \in [-\frac{1}{2}, \frac{1}{2}]$, for leaves.

Let $\mathcal{F}$ be a measured foliation of M. We say that $(M_0, \mathcal{F}_0)$ is a *normal form* of $\mathcal{F}$ if $(M_0, \mathcal{F}_0)$ is in normal form and $\mathcal{F}$ is obtained by enlargement of $(M_0, \mathcal{F}_0)$.

PROPOSITION 6.9. *Every measured foliation of M has a normal form.*

Proof. Let $(\mathcal{F}, \mu)$ be a measured foliation of M. Up to renumbering, we can suppose that

$$I(\mathcal{F}, \mu; [K_1]) \neq 0, \ldots, I(\mathcal{F}, \mu; [K_\ell]) \neq 0$$

and that $\quad I(\mathcal{F}, \mu; [K_{\ell+1}]) = \cdots = I(\mathcal{F}, \mu; [K_{3g-3}]) = 0.$

Then, by Proposition 5.9, by changing $\mathcal{F}$ in its class, we obtain that $\mathcal{F}$ is transverse to $K_i \times \{t\}$ for all $t \in [-1, 1]$ and $i = 1, \ldots, \ell$.

Let M' be the complement of the annuli $K_i \times (-1, 1)$, $i = 1, \ldots, \ell$. We have an induced measured foliation $(\mathcal{F}', \mu)$ that is transverse to the boundary. The curves K_i, $i \geq \ell + 1$ are contained in the interior of M'.

For $i \geq \ell + 1$, we have $I(\mathcal{F}', \mu; [K_i]) = 0$. Indeed, otherwise there exists $\mathcal{F}''$, a measured foliation of M' equivalent to $\mathcal{F}'$, coinciding with $\mathcal{F}'$ near the boundary, and transverse to K_i. But then

$$I(\mathcal{F}, \mu; [K_i]) = I(\mathcal{F}'', \mu; [K_i]) \neq 0,$$

which is a contradiction.

Applying Proposition 5.9 to $\mathcal{F}'$, M', and $K_{\ell+1}$, we obtain the following: up to changing $\mathcal{F}'$ in its class, there exists a homotopy $f_t : K_{\ell+1} \to M'$ such that $f_0 = (K_{\ell+1} \hookrightarrow M')$ and $f_1(K_{\ell+1})$ is an invariant set Σ_1 of $\mathcal{F}'$. When we apply the same proposition to $K_{\ell+2}$, we must do further Whitehead operations on $\mathcal{F}'$. These operations induce shifts on Σ_1, but $K_{\ell+1}$ continues to be homotopic onto an invariant set. Continuing in this way, we can eventually modify $\mathcal{F}'$ and find a homotopy $f_t : K_{\ell+1} \cup \cdots \cup K_{3g-3} \to M'$ so that f_0 is inclusion and the image of f_1 is an invariant set Σ of $\mathcal{F}'$.

Let $N \subset M'$ be a regular neighborhood of Σ (we mean regular with respect to the foliation); then $\mathcal{F}'$ is obtained by enlargement of a foliation $\mathcal{F}'_1$ on $M'_1 = \overline{M' - N}$. By construction, no component of N is a disk. As M'_1 has a measured foliation, no component of M_1' is a disk. We conclude that no component of $\partial M_1'$ is homotopic to a point.

We can suppose (by "engulfing") that the interior of N contains all the K_i, $i \geq \ell + 1$. Indeed, the singular map f_1 is close to an immersion f_1'. Then, if f_1' has double points, there exists a "Whitney disk" Δ through which one can do a homotopy whose effect is to decrease the number of double points of f_1'. But since no component of ∂N is homotopic to a point, if $\partial \Delta$ is contained in N, then Δ is contained in N. By induction on the number of double points, we see that one can deform f_1' to an embedding where the image is contained in N. By the work of Epstein (see Exposé 3), f_1' will be isotopic to f_0.

For any j, the intersection $R_j \cap M_1'$ is made of at most one pair of pants ("concentric to R_j") and of a certain number of annuli parallel to the components of the boundary of R_j. The boundary of a component of $R_j \cap M_1'$ can be of one of two types:

- $K_i \times \{\pm 1\}$ for $i \leq \ell$, that is, a transverse curve to $\mathcal{F}_1'$;

- A cycle of leaves of $\mathcal{F}_1'$, parallel to one of the K_i, where $i \geq \ell + 1$.

Then a component of $M_1' \cap R_j$ that is an annulus can only be foliated by circles. But a component of $M_1' \cap R_j$ that is a pair of pants might not carry a good foliation. We apply Proposition 6.8, which allows us to replace $(M_1', \mathcal{F}_1')$ by a foliation $(M_0', \mathcal{F}_0')$ satisfying:

- $\mathcal{F}_0'$ is equal to $\mathcal{F}_1'$ in a neighborhood of $\partial M'$.

- $(M_1', \mathcal{F}_1')$ is obtained by enlargement of $(M_0', \mathcal{F}_0')$.

- If V is a component of $R_j \cap M_0'$ that is a pair of pants, then $\mathcal{F}_0'|V$ is a good foliation.

By an obvious isotopy of $(M_0', \mathcal{F}_0')$, making, for example, the V above coincide with R_j', we obtain the desired normal form.   □

Definition of $\mathcal{NF}$. Two pairs $(M_0, \mathcal{F}_0)$ and $(M_0', \mathcal{F}_0')$ are *equivalent* if $M_0 = M_0'$ and $\mathcal{F}_0'$ can be obtained from $\mathcal{F}_0$ by a finite sequence of elementary operations of the following types:

- Whitehead operation with support in one of the R_j'.

- Isotopy with support in $K_i \times [-1, 1] \cap M_0$.

- Isotopy with support in R_j.

The set of equivalence classes is denoted $\mathcal{NF}$ (or $\mathcal{NF}(M)$). Enlargement induces a function $\mathcal{NF} \to \mathcal{MF}$, which is surjective by the preceding proposition. We will see later that it is in fact bijective.

The classification of foliations in normal form. We refer here to the permissible decomposition of M given in Section 6.4:

$$M = \left(\bigcup_{j=1}^{2g-2} R'_j\right) \cup \left(\bigcup_{i=1}^{3g-3} K_i \times [-1,1]\right).$$

For each R'_j, we choose once and for all a diffeomorphism to the standard pair of pants, respecting the orientation. Further, for each $i = 1, \ldots, 3g-3$, we choose curves K'_i and K''_i; if R_{j_1} and R_{j_2} are the two pairs of pants (distinct, because the decomposition is permissible) that contain K_i, then K'_i is an essential simple closed curve[2] that is not parallel to any of the curves of $\partial R_{j_1} \cup \partial R_{j_2}$. The curve K''_i is obtained from K'_i by a positive Dehn twist about K_i (see Figure 6.16).

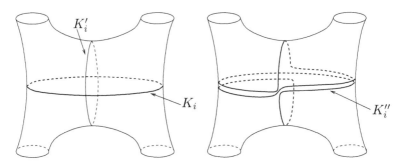

Figure 6.16.

We set

$$B = \{(m_i, s_i, t_i) : i = 1, \ldots, 3g-3, \ m_i, s_i, t_i \geq 0, \ (m_i, s_i, t_i) \in \partial(\nabla \leq)\}.$$

This is a cone in $\mathbb{R}_+^{9g-9}$ that is homeomorphic to $\mathbb{R}^{6g-6}$. We are going to construct a function

$$\mathcal{NF} \to B - 0.$$

Let $(M_0, \mathcal{F}_0)$ be a representative of an element of $\mathcal{NF}$. Without changing its class, we can suppose that $\mathcal{F}_0|_{R'_j}$ is in canonical form for all j such that $M_0 \cap R'_j \neq \emptyset$. The invariant m_i is the measure of K_i; it is zero if $K_i \cap M_0 = \emptyset$. The invariants s_i and t_i depend on the form of the induced foliation on the annulus $K_i \times [-1,1]$; there are three cases.

Case 1: $K_i \times [-1,1] \cap M_0 \subset K_i \times \{-1\} \cup K_i \times \{1\}$.

In this case, we set $s_i = t_i = 0$.

Case 2: $K_i \times [-\frac{1}{2}, \frac{1}{2}] = K_i \times (-1,1) \cap M_0$.

---

[2]If the two sides of K_i were to belong to the same pair of pants, then K'_i would not be embeddable.

In this case, the annulus is foliated by circles; then $s_i = t_i$ is the width of the annulus, that is to say the measure of a transversal.

Note that, in the two first cases, $m_i = 0$; thus it is clear that (m_i, s_i, t_i) belongs to $\partial(\nabla \leq)$.

Case 3: $M_0 \cap K_i \times [-1, 1] = K_i \times [-1, 1]$.

In this case, the foliation is transverse to the circles $K_i \times \{x\}$ for all $x \in [-1, 1]$, and M_0 contains the two pairs of pants R'_k and R'_l adjacent to $K_i \times [-1, 1]$. We have a pants seam J_k and a pants seam J_l. There exist then two arcs S_i and S'_i in $K_i \times [-1, 1]$ such that $J_k \cup S_i \cup J_l \cup S'_i$ is a closed curve homotopic to K'_i. The homotopy classes of S_i and S'_i, with endpoints fixed, are completely determined. We take S_i and S'_i to be arcs of minimal length. Given a fixed orientation of the surface, we can distinguish S_i from S'_i. We set

$$s_i = \mu_0(S_i) \quad \text{and} \quad s'_i = \mu_0(S'_i)$$

where μ_0 is the measure accompanying the foliation $\mathcal{F}_0$.

In the same way, we construct arcs T_i and T'_i such that $J_k \cup T_i \cup J_l \cup T'_i$ is homotopic to K''_i. We set

$$t_i = \mu_0(T_i) \quad \text{and} \quad t'_i = \mu_0(T'_i).$$

In short, the invariants (m_i, s_i, t_i) are in this case the invariants classifying the induced foliation on the annulus $K_i \times [-1, 1]$, in the sense of the classification of Section 6.1. In particular, we have: $(m_i, s_i, t_i) \in \partial(\nabla \leq)$.

It is very easy to see that the invariants m_i, s_i, and t_i only depend on the class of $(M_0, \mathcal{F}_0)$ in $\mathcal{NF}$.

LEMMA 6.10. *The image of $\mathcal{NF}$ in B does not contain 0.*

Proof. Let $(M_0, \mathcal{F}_0)$ be given. As M_0 is not empty, we have one of the following situations:

(1) For some i, $M_0 \cap K_i \times (-1, 1) = K_i \times [-\frac{1}{2}, \frac{1}{2}]$. In this case, we have $(s_i, t_i) \neq 0$.

(2) For some j, R'_j is contained in M_0. In this case, as the induced foliation is a good foliation, one of the curves of the boundary has nonzero measure.   □

Given what has been said about the classification of the measured foliations on the annulus and the pair of pants, we can leave as an exercise the details of the following proposition.

PROPOSITION 6.11. (Classification of measured foliations in normal form) *The function constructed above*

$$\mathcal{NF} \to B - 0$$

is a bijection.

6.5 CLASSIFICATION OF MEASURED FOLIATIONS

We consider here a closed orientable surface M of genus $g \geq 2$. We return to the other cases in Exposé 11.

Recall that a function f is positively homogeneous of degree 1 if $f(\lambda x) = \lambda f(x)$ whenever $\lambda > 0$.

PROPOSITION 6.12. *There exists a continuous function $\theta : I_*(\mathcal{MF}) \to B$ that is positively homogeneous of degree 1 and that makes the following diagram commutative:*

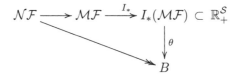

Proof. Since $\mathcal{NF} \to \mathcal{MF}$ is a surjection, it suffices to show that, for a foliation with normal form $(M_0, \mathcal{F}_0)$, the invariants m_i, s_i, and t_i depend only on the measures of simple closed curves.

It is immediately clear that $m_i = I(\mathcal{F}_0, \mu_0; [K_i])$. We will show in Appendix C that s_i and t_i are determined by $I(\mathcal{F}_0, \mu_0; [K_i'])$ and $I(\mathcal{F}_0, \mu_0; [K_i''])$, via homogeneous continuous formulas. $\square$

Since $\mathcal{NF} \to B$ is an injection, we immediately draw the following corollaries.

THEOREM 6.13. *Two measured foliations $(\mathcal{F}, \mu)$ and $(\mathcal{F}', \mu')$ on a surface M are Whitehead equivalent if and only if, for all simple curves γ of M, we have*

$$I(\mathcal{F}, \mu; [\gamma]) = I(\mathcal{F}', \mu'; [\gamma]).$$

PROPOSITION 6.14. *The enlargement function $\mathcal{NF} \to \mathcal{MF}$ is a bijection.*

Now, we can identify $\mathcal{MF}$ with its image via I_*, to provide $\mathcal{MF}$ with the topology induced by $\mathbb{R}_+^{\mathcal{S}}$ and to complete to $\overline{\mathcal{MF}} = \mathcal{MF} \cup 0$.

THEOREM 6.15. *The function θ is a homeomorphism of $\overline{\mathcal{MF}}$ onto $B \cong \mathbb{R}^{6g-6}$, and is positively homogeneous of degree 1. Consequently, $\mathcal{PMF}$ (the space of projective measured foliations, see Section 6.7) is homeomorphic to S^{6g-7}.*

We already know that the classifying function θ is a continuous bijection. If one shows that $\mathcal{MF}$ is a topological manifold, then invariance of domain implies that θ is also open, and the theorem will follow.

To prove that $\mathcal{MF}$, with the topology of $\mathbb{R}_+^{\mathcal{S}}$, is a topological manifold, we use the following lemmas.

LEMMA 6.16 (Change of decomposition). *Let $\mathcal{K}$ be a permissible decomposition of M into pairs of pants and $(M_0, \mathcal{F}_0, \mu_0)$ a measured foliation in normal form with respect to $\mathcal{K}$. There exists another permissible decomposition $\widehat{\mathcal{K}} = \{\widehat{K}_1, \ldots, \widehat{K}_{3g-3}\}$, so that $\widehat{m}_i = I(\mathcal{F}_0, \mu_0; [\widehat{K}_i])$ is nonzero for each i.*

N.B. It is not said that $(\mathcal{F}_0, \mu_0)$ is in normal form with respect to this decomposition.

Proof. We suppose at first that, for all i, we have

$$I(\mathcal{F}_0, \mu_0; [K_i]) = 0.$$

In particular, the support of M_0 is concentrated in the annuli $K_i \times [-\frac{1}{2}, \frac{1}{2}]$. We look at one such i and the two pairs of pants R'_k and R'_ℓ that intersect $K_i \times [-1, 1]$ (Figure 6.17).

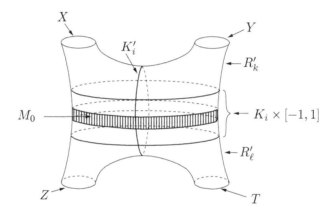

Figure 6.17.

If we are in the situation suggested by the figure, where neither the pair (X, Z) nor the pair (Y, T) bounds an annulus, we replace K_i by K'_i; this gives a permissible decomposition where $I(\mathcal{F}_0, \mu_0; [K'_i]) \neq 0$. Otherwise, if (X, Z) bounds an annulus, we construct the simple curve K'''_i, which is obtained from K'_i by a half twist along K_i and we replace K_i by K'''_i (Figure 6.18). The resulting decomposition is permissible because the pair (Y, Z) does not bound an annulus (otherwise (Y, X) would bound an annulus and $\mathcal{K}$ would not be permissible).

We are reduced to the situation where at least one of the m_i is nonzero, and, considering any fixed i:

- If $m_i = 0$, then K_i avoids M_0 or is a cycle of leaves of $\mathcal{F}_0$.

- If $m_i \neq 0$, then $K_i \cap M_0$ is transverse to $\mathcal{F}_0$.

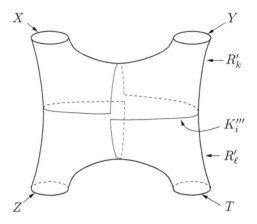

Figure 6.18.

Let us say then that we have a pair of pants R, bounded by $K_1 \cup K_2 \cup K_3$, with $m_1 = 0$ and $m_2 \neq 0$. For the enlargement of the foliation induced on R, we have the three possibilities of Figure 6.19.

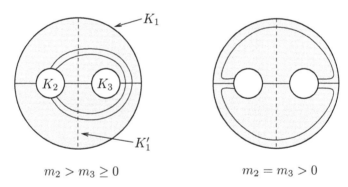

$m_2 > m_3 \geq 0$ $m_2 = m_3 > 0$

(by switching the letters K_2
and K_3, we obtain the case
$m_3 > m_2 \geq 0$)

Figure 6.19.

As before, we construct K_1' (or K_1''') which is transverse to $\mathcal{F}_0$ and which gives a new permissible decomposition where $m_1 \neq 0$ (use the dashed arc in the figure).
□

LEMMA 6.17. *Let $(\mathcal{F}_0, \mu_0)$ be a measured foliation in regular position with respect to a permissible decomposition $\mathcal{K}$. We suppose that, for all $i = 1, \ldots, 3g-3$, we have*

$$m_i^0 = m_i(\mathcal{F}_0, \mu_0) \neq 0.$$

Then the function $\theta^{-1} : B - 0 \to I_*(\mathcal{MF}) \subset \mathbb{R}_+^{\mathcal{S}}$ is continuous at the point with coordinates $(m_i^0, s_i^0, t_i^0)_{i=1,\ldots,3g-3}$.

Remark 1. This proves that $I_*(\mathcal{MF})$ is a topological manifold in a neighborhood of $(\mathcal{F}_0, \mu_0)$; therefore, if we apply Lemma 6.16, $I_*(\mathcal{MF})$ is a topological manifold globally.

Remark 2. Lemma 6.17 would be trivial if one could lay out explicit formulas that, for all $\gamma \in \mathcal{S}$, express $I(\mathcal{F}, \mu; \gamma)$ as a function of $(m, s, t)(\mathcal{F}, \mu)$ and of $(m, s, t)(\gamma)$.

Proof. We denote by E the set of measured foliations transverse to all the curves K_i of the decomposition $\mathcal{K}$, without the equivalence relation. Let

$$B^0 = \{(m_i, s_i, t_i)_i \in B : m_i \neq 0, \text{ for all } i = 1, \ldots, 3g - 3\}.$$

There exists a section of θ, call it $\sigma : B^0 \to E$, with the following properties:

(1) A foliation in the image of σ is in normal form with respect to $\mathcal{K}$ and, for all i, $\mathcal{F}|_{K_i \times [-1,1]}$ varies continuously in the sense of the topology of 1-forms.

(2) If α is an arc of R'_j that connects the boundary to the boundary and that is transverse to $(\mathcal{F}_0, \mu_0) = \sigma((m_i^0, s_i^0, t_i^0)_i)$, then α is transverse to $(\mathcal{F}, \mu) = \sigma((m_i, s_i, t_i)_i)$, for (m_i, s_i, t_i) close enough to (m_i^0, s_i^0, t_i^0); further, $\mu(\alpha)$ varies continuously.

(3) Let $\alpha_0 * \beta_0$ be an arc of R'_j, going from the boundary to the boundary, where α_0 is transverse to $\mathcal{F}_0$, where β_0 is in a leaf and where $\alpha_0 * \beta_0$ is quasitransverse to $\mathcal{F}_0$. Then, for (m_i, s_i, t_i) close enough to (m_i^0, s_i^0, t_i^0), there exists an arc $\alpha * \beta \subset R'_j$ going from the boundary to the boundary such that:

(a) $\alpha * \beta$ is C^0-close to $\alpha_0 * \beta_0$.

(b) $\alpha \pitchfork \mathcal{F}$, where $(\mathcal{F}, \mu) = \sigma((m_i, s_i, t_i)_i)$.

(c) $\alpha * \beta$ is quasitransverse to $\mathcal{F}$.

(d) $\mu(\beta)$ and $|\mu(\alpha * \beta) - \mu_0(\alpha_0)|$ are small.

(4) Same condition for arcs of the form $\beta_0 * \alpha_0 * \beta'_0$.

[In a certain sense, these conditions say that σ is continuous. But is there a good topology on E?]

We will be satisfied with a brief outline for the existence of σ. Since we define σ only on B_0, we will only use in each pair of pants the models (1), (2), and (3) of Section 6.2. As long as we stay in the interior of the fundamental triangle, we can "continuously" vary the actual realizations of these models as well as the corresponding pants seams. This makes it possible to reglue the pieces in order

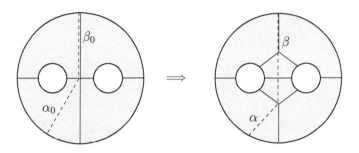

Figure 6.20. In a neighborhood of "$m_1 = m_2 + m_3$"

to obtain a section σ that is continuous in the topology of vector fields (outside of the singularities). Figure 6.20 illustrates the third condition.

Given σ, we can easily finish the proof of Lemma 6.17. Let $\gamma \in \mathcal{S}$ and let σ_γ be the corresponding component of σ:

$$\sigma_\gamma : B_0 \to \mathbb{R}_+.$$

We want to show that this function is continuous in $(m_i^0, s_i^0, t_i^0)_i$. As we remarked after the statement of Proposition 5.9, we can find an immersion γ_0' that is quasitransverse to $\mathcal{F}_0$, that is the limit of embeddings, and that is homotopic to γ. Let us say that

$$\gamma_0' = \alpha_1^0 * \beta_1^0 * \alpha_2^0 \ldots,$$

where α_i^0 is transverse to $\mathcal{F}_0$ and where β_i^0 is contained in the leaves (and singular points). We remark right away that the μ-length of the representative γ_0' varies continuously. It follows that σ_γ is upper semicontinuous (this observation is not logically needed).

By properties (3) and (4) of σ, we construct for (m_i, s_i, t_i) close to (m_i^0, s_i^0, t_i^0) another immersed curve

$$\gamma' = \alpha_1 * \beta_1 * \alpha_2 * \cdots$$

that is homotopic to γ_0' and that satisfies:

1. α_i and β_i are glued quasitransversally to $\mathcal{F}$, where $(\mathcal{F}, \mu) = \sigma((m_i, s_i, t_i)_i)$.
2. α_i is transverse to $\mathcal{F}$ and $\mu(\alpha_i)$ is close to $\mu_0(\alpha_i^0)$.
3. $\mu(\beta_i)$ is small.

With endpoints fixed, β_i is isotopic to $\overline{\beta_i}$, which is quasitransverse to $\mathcal{F}$. We have $\mu(\overline{\beta_i}) \leq \mu(\beta_i)$. Using property (3) of σ, we easily see that $\overline{\gamma'} = \alpha_1 * \overline{\beta_1} * \alpha_2 * \cdots$, which is piecewise quasitransverse to $\mathcal{F}$, is really globally quasitransverse to $\mathcal{F}$. We therefore have

$$I(\mathcal{F}, \mu; [\gamma]) = \mu(\alpha_1) + \mu(\overline{\beta_1}) + \cdots,$$

a sum that, term by term, is close to

$$I(\mathcal{F}_0, \mu_0; [\gamma]) = \mu_0(\alpha_1^0) + \mu(\beta_1^0) + \cdots = \sum_i \mu_0(\alpha_i^0).$$

$\square$

6.6 ENLARGED CURVES AS FUNCTIONALS

We have the following commutative diagram (cf. Section 5.4):

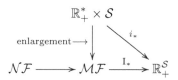

The arrow $\mathbb{R}_+^* \times \mathcal{S} \to \mathcal{MF}$ naturally factors through $\mathcal{NF}$. Indeed, if we represent an element of $\mathcal{S}$ by a curve γ having a minimal intersection with each K_j, then a partial enlargement of γ gives a foliation in normal form, as one sees by looking at each pair of pants.

Using the function $\theta : I_*(\mathcal{MF}) \to B$ of Proposition 6.12, we thus obtain

$$\overline{\Phi} : \mathbb{R}_+^* \times \mathcal{S} \ (\text{resp. } \mathcal{S}') \to B,$$

which to $\beta \in \mathcal{S}'$ associates $\{(\overline{m}_j(\beta), \overline{s}_j(\beta), \overline{t}_j(\beta)) : j = 1, \ldots, 3g - 3\}$. We recall that, in Exposé 4, for $\beta \in \mathcal{S}'$, we defined $\Phi(\beta) = \{(m_j(\beta), s_j(\beta), t_j(\beta))\}$. Unfortunately, $\overline{\Phi}(\beta)$ does not coincide with $\Phi(\beta)$. It is true that $m_j(\beta) = \overline{m}_j(\beta) = i(\beta, K_j)$, but the other coordinates differ because the pants seam is not chosen in the same way in the theory of curves as in the theory of foliations. Moreover, $\overline{\Phi}(\beta)$ does not always have integer coordinates.

To discuss this difference between the pants seams in the two theories, one must again examine the models on the standard pair of pants P^2. We observe that the pants seam for a multi-arc, associated to $\partial_1 P^2$, always coincides with that of the foliation obtained by enlargement, except if

$$m_1 > m_2 + m_3. \tag{6.1}$$

On the other hand, the pants seam from the theory of curves is appropriate for foliations. Evidently, the length of the associated arc A is only given by the formula in Section 6.3 if the inequality 6.1 is not satisfied. Otherwise we take

$$\text{length } A = m_2.$$

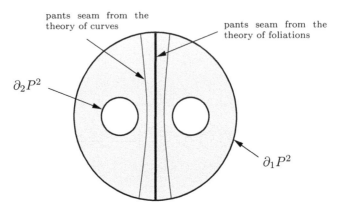

Figure 6.21.

Reflecting this change through the formulas of Appendix C, we obtain a new classification of foliations, via a homeomorphism

$$\theta_C : \overline{I_*(\mathcal{MF})} \to B,$$

which, this time, makes the following diagram commutative:

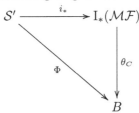

Therefore, $i_*(\mathcal{S}')$ is a "lattice" in $I_*(\mathcal{MF})$. As we know that $i_*(\mathcal{S}')$ is contained in $\overline{i_*(\mathbb{R}_+ \times \mathcal{S})}$, we see that $i_*(\mathbb{R}_+^* \times \mathcal{S})$ is dense in $I_*(\mathcal{MF})$. We have therefore demonstrated at the same time Theorem 4.10 and Proposition 6.18.

PROPOSITION 6.18. *In* $P(\mathbb{R}_+^{\mathcal{S}})$, *the set* $\pi \circ i_*(\mathcal{S}) = \mathcal{S}$ *is dense in* $\pi \circ I_*(\mathcal{MF})$. *Therefore,* $I_*(\mathcal{MF}) \cup \{0\} = \overline{i_*(\mathbb{R}_+ \times \mathcal{S})}$.

6.7 MINIMALITY OF THE ACTION OF THE MAPPING CLASS GROUP ON $\mathcal{PMF}$

Let M be a compact connected orientable surface without boundary, of genus ≥ 1. We always denote by π the projection $\mathbb{R}_+^{\mathcal{S}} - \{0\} \to P(\mathbb{R}_+^{\mathcal{S}})$, and by $\mathcal{PMF}$ the image of $\mathcal{MF}$ under π. The natural action of the *mapping class group* $\pi_0(\mathrm{Diff}(M))$ on $\mathcal{MF}$ gives, by passage to the quotient, a natural action of $\pi_0(\mathrm{Diff}(M))$ on $\mathcal{PMF}$.

The goal of this section is to show the following theorem.

THEOREM 6.19. *The action of $\pi_0(\mathrm{Diff}(M))$ on $\mathcal{PMF}$ is minimal.*

We recall that the action of a group on a topological space is called *minimal* if the orbit of each point is dense.

If α is a simple curve in M, we denote by $t_\alpha : M \to M$ a Dehn twist about α.

PROPOSITION 6.20. *Let α be a simple curve and $\mathcal{F}$ a measured foliation. For all curves β and for all integers $n \geq 0$, we have the inequality*

$$|I(t_\alpha^n(\mathcal{F}), [\beta]) - n\, I(\mathcal{F}, [\alpha]) i([\beta], [\alpha])| \leq I(\mathcal{F}, [\beta]).$$

Proof. If $\mathcal{F}$ is a foliation defined by a curve, the proposition is a particular case of Proposition A.1. Considering that the inequality is homogeneous in $\mathcal{F}$, the proposition is again true for $\mathcal{F}$ in $i_*(\mathbb{R}_+^* \times \mathcal{S})$. As $i_*(\mathbb{R}_+^* \times \mathcal{S})$ is dense in $\mathcal{MF}$, the inequality is true for every foliation $\mathcal{F}$. $\square$

COROLLARY 6.21. *Let $\mathcal{F}$ be a measured foliation and α a curve such that $I(\mathcal{F}, [\alpha]) \neq 0$. We have*

$$\lim_{n \to \infty} \pi(t_\alpha^n(\mathcal{F})) = \pi([\alpha]).$$

Proof. As a consequence of the preceding proposition, we have

$$\lim_{n \to \infty} \frac{1}{n\, i(\alpha, \mathcal{F})}\, t_\alpha^n(\mathcal{F}) = [\alpha] \quad \text{in } \mathcal{MF}.$$

$\square$

We prove the following particular case of Theorem 6.19.

LEMMA 6.22. *If γ is a curve that does not separate M, the orbit of $\pi([\gamma])$ under $\pi_0(\mathrm{Diff}(M))$ is dense in $\mathcal{PMF}$.*

Proof. We begin by remarking that the orbit of γ under $\pi_0(\mathrm{Diff}(M))$ consists of the (isotopy classes of) curves that do not separate M. Since $\mathcal{S}$ is dense in $\mathcal{PMF}$, it suffices to show that the closure of the orbit of γ contains also the curves that separate M. Let $\overline{\gamma}$ be such a curve. We can find a curve γ' that does not separate M and such that $i(\gamma', \overline{\gamma}) \neq 0$. By Corollary 6.21, we have $\lim_{n \to \infty} t_{\overline{\gamma}}^n(\gamma') = \overline{\gamma}$ in $\mathcal{PMF}$. Thus $\overline{\gamma}$ is in the closure of the orbit of γ' and also in that of γ, since these two orbits are the same. $\square$

Finally, we prove the theorem.

Proof of Theorem 6.19. Let $\mathcal{F}$ be a measured foliation. We can find a curve γ that does not separate M and such that $I(\mathcal{F}, \gamma) \neq 0$. By Corollary 6.21, the closure of the orbit of $\mathcal{F}$ in $\mathcal{PMF}$ contains γ, and thus also the orbit of γ. It follows from Lemma 6.22 that the orbit of $\mathcal{F}$ is dense in $\mathcal{PMF}$. $\square$

6.8   COMPLEMENTARY MEASURED FOLIATIONS

By definition a *complement* of a measured foliation $(\mathcal{F}, \mu)$ is a measured foliation $(\mathcal{F}', \mu')$ that is transverse to $(\mathcal{F}, \mu)$ (see Section 1.5).

PROPOSITION 6.23. *If* $(\mathcal{F}, \mu)$ *is a measured foliation, then there exists* $(\mathcal{F}'', \mu'') \overset{m}{\sim} (\mathcal{F}, \mu)$ *such that* $(\mathcal{F}'', \mu'')$ *admits a complement.*

Proof. By the results of Section 6.5, we obtain a foliation $(\mathcal{F}'', \mu'')$ that is equivalent to $(\mathcal{F}, \mu)$, and a pair of pants decomposition such that for all j, we have $i(\mathcal{F}'', K_j) \neq 0$.

By enlarging the multicurve provided by the K_j, we obtain the desired $\mathcal{F}'$.
□

Remark. This result is equivalent to the theorem of Hubbard–Masur ([HM79]) and Kerckhoff ([Ker80]), which states that $\mathcal{MF}(M^2)$ is realized by the holomorphic quadratic differentials on M^2. We refer to [DV76] and [HM79] for details on the relationship between quadratic differentials and measured foliations.

Appendix C

Explicit Formulas for Measured Foliations

by Albert Fathi

On the "double pair of pants," or sphere with four holes, we consider the curves K, K', and K'' (see Figure C.1).

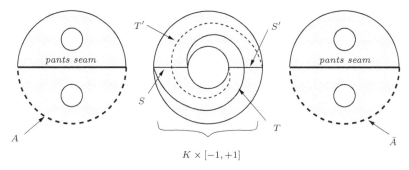

$$K \times [-1, +1]$$

$K' \simeq$ pants seam $\cup\, S \cup$ pants seam $\cup\, S'$

$K'' \simeq$ pants seam $\cup\, T \cup$ pants seam $\cup\, T'$

A and $\bar{A}$ are on the same side with respect to K'

Figure C.1. The curves K, K', and K''

For a foliation in normal form with respect to this decomposition, we have defined three numbers (m, s, t), in addition to the four measures of the curves of the boundary (see Section 6.4).

PROPOSITION C.1. *There exist continuous formulas, positively homogeneous of degree 1, giving s and t as functions of the minimal measures of the isotopy classes $[K]$, $[K']$, and $[K'']$ and of the curves of the boundary of the double pair of pants.*

Proof. We use the following notation: m is the length of K, and s, t, s', t', a and $\bar{a}$ are the lengths of the arcs S, T, S', T', A, and $\bar{A}$, defined in Section 6.4 and recalled in Figure C.1.

Claim 1. If $m \neq 0$, we can calculate s and t as functions of $\alpha = s + s'$, $\beta = t + t'$, m, a, and $\bar{a}$.

We trivialize the annulus $K \times [-1, 1]$ in such a way that the projection onto K foliates like the given foliation. In the covering $\mathbb{R} \times [-1, 1]$ of the annulus, the covering group acts as translation by m; we have a picture like Figure C.2, where we have drawn the (line) segments realizing the minimum lengths of the arcs

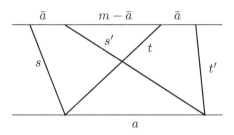

Figure C.2.

lifting S, T, etc. Obvious geometric reasons imply that the upper endpoints of these arcs always appear in the indicated order. We also recall something visible in the figure:

$$(m, s, t) \in \partial(\nabla \leq), \qquad (m, s', t') \in \partial(\nabla \leq).$$

From this it follows that $(2m, \alpha, \beta) \in (\nabla \leq)$. Thus we have

$$
\begin{aligned}
\alpha &\leq \beta + 2m, \\
\beta &\leq \alpha + 2m, \\
2m &\leq \alpha + \beta,
\end{aligned}
$$

and, of course, $m \geq a$, and $m \geq \bar{a}$. Moreover, $(s, s', \bar{a}, a)$ and $(t, t', \bar{a}, a)$ are the lengths of the sides of degenerate quadrilaterals; thus we have

$$\alpha, \beta \geq |a - \bar{a}|.$$

We describe the possible configurations in terms of the angle that each arc makes with the horizontal in the universal cover. We exclude some configurations by remarking that if S makes an angle less than or equal to $\pi/2$, then T' cannot make an angle greater than $\pi/2$; otherwise we would have $a > m$.

Configuration I is characterized by $\beta = \alpha + 2m$; further, we have

$$\beta \geq \alpha,$$

$$
\begin{cases}
s + s' = \alpha, \\
s - s' = a - \bar{a}, \\
t = s + m, \\
t' = s' + m.
\end{cases}
$$

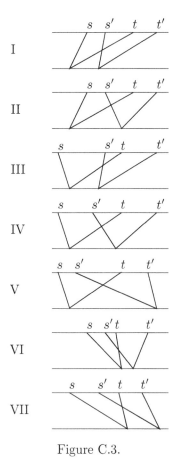

Figure C.3.

Indeed, taking into account that $(m, s, t) \in \partial(\nabla \leq)$ and $(m', s', t') \in \partial(\nabla \leq)$, we see that $\beta = \alpha + 2m$ implies $t = s + m$ and $t' = s' + m$, which determines configuration I.

Configuration II is characterized by $\alpha = a - \bar{a}$; further, we have

$$\beta \geq \alpha,$$

$$\begin{cases} s + s' = \alpha, \\ s - s' = \beta - 2m, \\ t = s + m, \\ t' = m - s'. \end{cases}$$

Indeed, $s + s' + \bar{a} = a$ determines ; as $a < m$, the angle of T' must be smaller than $\pi/2$.

Analogous reasoning allows one to establish characterizations of the other cases.

Configuration III is characterized by $\alpha = \bar{a} - a$; further, we have

$$\beta \geq \alpha,$$

$$\begin{cases} s + s' = \alpha, \\ s - s' = 2m - \beta, \\ t = m - s, \\ t' = s' + m. \end{cases}$$

Configuration IV is characterized by $\alpha + \beta = 2m$; further, we have

$$\begin{cases} s + s' = \alpha, \\ s - s' = \bar{a} - a, \\ t = m - s, \\ t' = m - s'. \end{cases}$$

Configuration V is characterized by $\beta = a - \bar{a}$; further, we have

$$\alpha \geq \beta,$$

$$\begin{cases} s + s' = \alpha, \\ s - s' = \bar{a} - a, \\ t = m - s, \\ t' = m - s'. \end{cases}$$

Configuration VI is characterized by $\beta = \bar{a} - a$; further, we have:

$$\alpha \geq \beta,$$

$$\begin{cases} s + s' = \alpha, \\ s - s' = \bar{a} - a, \\ t = s - m, \\ t' = m - s'. \end{cases}$$

Configuration VII is characterized by $\alpha = \beta + 2m$; further, we have

$$\alpha \geq \beta,$$

$$\begin{cases} s + s' = \alpha, \\ s - s' = \bar{a} - a, \\ t = s - m, \\ t' = s' - m. \end{cases} \qquad \square$$

By a small calculation, we see that in cases I, II, III, and IV we have:

$$\begin{cases} s = |m + \frac{\bar{a} - a - \beta}{2}|, \\ t = \frac{a - \bar{a} + \beta}{2}, \end{cases} \tag{C.1}$$

and that in cases IV, V, VI, and VII, we have:

$$\begin{cases} s = \frac{\alpha + \bar{a} - a}{2}, \\ t = |m + \frac{a - \bar{a} - \alpha}{2}|. \end{cases} \tag{C.2}$$

We introduce a closed positive cone in $\mathbb{R}_+^5$:

$$\mathcal{C} = \{(\alpha, \beta, m, a, \bar{a}) \in \mathbb{R}_+^5 | (\alpha, \beta, 2m) \in (\nabla \leq), m \geq a, m \geq \bar{a},$$
$$\alpha \geq |a - \bar{a}|, \beta \geq |a - \bar{a}|; \text{one of the following equalities is satisfied:}$$
$$\alpha = |a - \bar{a}|, \beta = |\bar{a} - a|, \alpha = \beta + 2m, \alpha + \beta = 2m, \beta = \alpha + 2m\}.$$

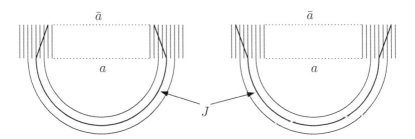

Figure C.4.

We see that $\alpha = |a - \bar{a}|$. By analyzing in an analogous manner what happens with $\bar{J}$, we obtain the proof of the claim. $\qquad \square$

If $m = 0$, the preceding formulas become $k' = \alpha + j + \bar{j}$ and $k'' = \beta + j + \bar{j}$. If we look at the models, we see that they agree on the level of the geometry. For this observation, do not forget the case where one of the pants is not in the support of the foliation. In this case, the three measures of the boundary, as well as the length of the pants seam, are zero.

Fundamental remark. This appendix is universal! Precisely, we can change the pants seam for each type of foliation of the standard pair of pants to any other arc that has the following properties:

1. It stays in the same isotopy class.

2. It realizes the minimum transverse length in this class.

A new choice of arcs on the models leads to a new classifying homeomorphism $\theta : I_*(\mathcal{MF}) \to B - \{0\}$. This will be built from the formulas of this appendix, which stay exactly the same. The only change is in the expression of the length of the arc A associated to each pants seam.

We define $\phi : \mathcal{C} \to \mathbb{R}^2$ by the formulas C.1 if $\beta \geq \alpha$ and by the formulas C.2 if $\beta \leq \alpha$. It is easy to see that the two formulas coincide if $\alpha = \beta$. On the other hand, if (s,t) are the coordinates of ϕ, we see that (s,t) belongs to $\mathbb{R}_+^2$ and that (m, s, t) belongs to $\partial(\nabla \leq)$.

The interest in introducing $\mathcal{C}$ is to show that the function θ extends to a closed subcone of $\mathbb{R}_+^S$.

We remark that if $m = 0$ (and as a consequence, $a = 0$, $\bar{a} = 0$), we obtain for the above formulas

$$s = t = \frac{\alpha}{2} = \frac{\beta}{2},$$

which coincides with what the geometry says.

We set $k' = I(\mathcal{F}, \mu; [K'])$, $k'' = I(\mathcal{F}, \mu; [K''])$, j and $\bar{j}$ the lengths of the pants seams J and $\bar{J}$ of the pairs of pants containing A and $\bar{A}$, respectively.

Claim 2. If $m \neq 0$, we have $\alpha = \sup(|a - \bar{a}|, k' - j - \bar{j})$ and $\beta = \sup(|a - \bar{a}|, k'' - j - \bar{j})$.

First of all, by definition of k', we have $\alpha + j + \bar{j} \geq k'$, and we have already seen that $\alpha \geq |a - \bar{a}|$. If J and $\bar{J}$ are nonzero lengths (which means that the chosen arcs pass through singularities), we easily replace $J \cup S \cup \bar{J} \cup S'$ with a quasitransverse curve of the same length; in this case, $k' = \alpha + j + \bar{j}$. If J is of zero length (piece of a smooth leaf) and if S and S' leave from different sides of J, we replace $S \cup J \cup S'$ with a transversal of the same measure (Figure C.5).

Figure C.5.

If S and S' leave from the same side of J, we have one of the two configurations of Figure C.4.

Exposé Seven

Teichmüller Space

by Adrien Douady; notes by François Laudenbach

Let M be a compact surface with negative Euler characteristic $\chi(M)$. We consider the space $\mathcal{H}$ of metrics on M, where the curvature is -1, and where the boundary of M is geodesic. This space is nonempty and is endowed with the C^∞ topology for contravariant tensor fields. The group $\text{Diff}_0(M)$—the group of diffeomorphisms of M isotopic to the identity, equipped with the C^∞ topology—acts on $\mathcal{H}$ on the left by pullback: if $m \in \mathcal{H}$ and $\phi \in \text{Diff}_0(M)$, then $\phi \cdot m = \phi^* m \in \mathcal{H}$. The quotient space $\mathcal{T} = \mathcal{H}/\text{Diff}_0(M)$ is the *Teichmüller space* of M. When M is orientable, this definition coincides with the classical definition as the space of complex structures up to isotopy, by the uniformization theorem [Spr57]. It is known that this space is homeomorphic to a cell [FK65]. Earle and Eells have shown that $\mathcal{H}$ is the total space of a principal bundle over Teichmüller space [EE69].

The program here is to establish a parametrization of Teichmüller space that depends only on the lengths of simple closed geodesics.

Recall that $\mathcal{S}$ is the set of isotopy classes of simple closed curves that are not homotopic to a point in M. If m is a hyperbolic metric, and $\alpha \in \mathcal{S}$, then $\ell(m, \alpha)$ is the length of the unique geodesic in the isotopy class α. We thus have a map

$$\ell_* : \mathcal{T} \to \mathbb{R}_+^{\mathcal{S}}$$

given by the formula $\langle \ell_*(m), \alpha \rangle = \ell(m, \alpha)$.

PROPOSITION 7.1. *For a fixed $\alpha \in \mathcal{S}$, the map that associates to $m \in \mathcal{H}$ the m-geodesic in the class α is continuous in the C^∞ topology.*

COROLLARY 7.2. *The map ℓ_* is continuous.*

One way to prove the proposition is to use the convexity of the "displacement function" (a theorem of Bishop–O'Neill [BO69]; see the paper of Bourguignon [Bou71]). We give a different proof.

Proof of Proposition 7.1. We denote by Γ the set of pairs (m, γ) where m is a hyperbolic metric and $\gamma \colon S^1 \to M$ is a constant-speed parametrization of the m-geodesic of α. We give Γ the topology induced from the C^∞ topology on the product space

$$\mathcal{H} \times C^\infty(S^1, M).$$

We consider the projection $p\colon \Gamma \to \mathcal{H}$ onto the first factor. We wish to show that p is proper.

We let TM denote the tangent bundle of M, and we consider the subset of $\mathcal{H} \times TM$ given by

$$C = \{(m, v) \mid \forall\, t,\ \exp_m(t+1)v = \exp_m tv$$

and the closed curve $t \in [0, 1] \to \exp_m tv$ is in the class $\alpha\}$.

In the product topology on $\mathcal{H} \times TM$, the set C is closed. If S^1 is obtained by identifying the endpoints of $[0, 1]$, one has an obvious map $C \to \Gamma$ which is surjective; by the theory of differential equations, it is continuous. The properness of p follows from the properness of the projection $q\colon C \to \mathcal{H}$, as we shall prove.

We know that $m \in \mathcal{H} \mapsto \ell(m, \alpha)$ is an upper semicontinuous function. Hence if m belongs to a compact set K, the set

$$\{\ell(m, \alpha) \mid m \in K\}$$

is bounded. Let $(m, v) \in q^{-1}(K)$; the quantity

$$\sqrt{m(v, v)} = \ell(m, \alpha)$$

is then bounded. Let $m_0 \in K$; there exists $\lambda > 0$ such that, for all $w \in TM$, and all $m \in K$, one has

$$m_0(w, w) \leq \lambda\, m(w, w).$$

Thus, if $(m, v) \in q^{-1}(K)$, then $m_0(v, v)$ is bounded. Finally, $q^{-1}(K)$ is compact since it is closed in a product of compact sets.

The group $O(2)$ of rotations acts naturally on Γ: for $r \in O(2)$, $(m, \gamma) * r = (m, \gamma \circ r)$. The quotient is the space of m-geodesics of α, for $m \in \mathcal{H}$. Since m has negative curvature, we have that p induces a bijection $\Gamma/O(2) \to \mathcal{H}$, which is continuous and proper by the above. Since the spaces considered are metrizable, the inverse is also continuous. □

From now on, to simplify the exposition, we suppose that M is a closed surface of genus g. We fix a decomposition $\mathcal{K}$ of M into pairs of pants R_i, $i = 1, \ldots, 2g - 2$, bounded by curves K_j, where $j = 1, \ldots, 3g - 3$. Each pair of pants is given with a parametrization onto some model, and every curve K_j is given with an orientation. We have a continuous map

$$L\colon \mathcal{T} \to (\mathbb{R}_+^*)^{3g-3}$$

defined by $L(m) = (\ell(m, K_i); i = 1, \ldots, 3g - 3)$, where m is a hyperbolic metric making the K_i geodesic (a so-called metric *adapted* to the decomposition).

Remark. From now on, $\mathcal{H}$ denotes the space of metrics adapted to $\mathcal{K}$. One sees easily that $\mathcal{T}$ is in bijection with the quotient of $\mathcal{H}$ by $\mathrm{Diff}(M, \mathcal{K}) \cap \mathrm{Diff}_0(M)$. To see that the topology is the same, we use Proposition 7.1 and the fact that the action of $\mathrm{Diff}(M)$ on the space of simple curves admits local sections [Pal60].

The set of "twists" along the curves K_i defines a continuous action θ of $\mathbb{R}^{3g-3}$ on $\mathcal{T}$. More precisely, let $K_i \times [0,1]$ be a collar of $K_i = K_i \times \{0\}$, given once and for all; the collars are assumed to be pairwise disjoint. Being given an adapted hyperbolic metric m and a number α, there exists a diffeomorphism $\phi_i(m,\alpha)$ of the collar $K_i \times [0,1]$ with the following properties:

1. $\phi_i(m,\alpha)$ is the identity on a neighborhood of $K_i \times \{1\}$.

2. $\phi_i(m,\alpha)$ is an isometry of m in a neighborhood of $K_i \times \{0\}$.

3. The lift of $\phi_i(m,\alpha)$ to the universal covering $\mathbb{R} \times [0,1]$ that is the identity on $\mathbb{R} \times \{1\}$ is a translation of distance $\alpha \, \ell(m, K_i)$ on $\mathbb{R} \times \{0\}$ in the direction indicated by the sign of α (the universal cover is given the lifted metric).

The twisted metric $\theta_i(m,\alpha)$ is defined by $\theta_i(m,\alpha) = \phi_i^*(m,\alpha)\,m$ for points of the collar $K_i \times [0,1]$ and by $\theta_i(m,\alpha) = m$ elsewhere.

For $(\alpha_1, \ldots, \alpha_{3g-3}) \in \mathbb{R}^{3g-3}$, let $\theta(m, \alpha_1, \ldots, \alpha_{3g-3})$ be the metric defined by $\theta_i(m,\alpha_i)$ in $K_i \times [0,1]$, and by m elsewhere. As the metric is adapted, its isotopy class is well-defined.

Remark 1. By the classification of hyperbolic metrics on pairs of pants (Exposé 3), the orbits of the action θ coincide exactly with the fibers of L. Corollary 7.4 below implies that this action is free.

Remark 2. The Dehn twist ρ along K_i, which is a global diffeomorphism of the surface with support in a collar of K_i, is an isometry (up to isotopy) of the metric $\theta_i(m,1)$ onto m. One therefore has, for all curves K',

$$\ell(\theta_i(m,1), [K']) = \ell(m, \rho([K'])).$$

Let R and R' be the two pairs of pants adjacent to K_i, and suppose that R contains the collar $K_i \times [0,1]$. Let K_i' be a simple curve in $R \cup R'$ intersecting K_i in two essential points (by this we mean that K_i' is not isotopic to a curve disjoint from K_i)—compare with Section 6.4. We denote by K_i'' the curve in $R \cup R'$ obtained from K_i' by a Dehn twist along K_i: $\rho(K_i') = K_i''$.

PROPOSITION 7.3. *The length $\ell(\theta_i(m,\alpha), [K_i''])$ is a strictly convex function of α that takes a minimum.*

We will prove the proposition after giving a corollary and two lemmas.

COROLLARY 7.4. *(1) Being given a metric m_0, there exists an isotopy class γ_i in $R \cup R'$ such that the function*

$$\alpha \mapsto \ell(\theta_i(m_0,\alpha), \gamma_i)$$

is strictly increasing for $\alpha > 0$.

(2) The length $\ell(\theta_i(m,\alpha), [K_i'])$ tends uniformly to ∞ as α tends to ∞ or to $-\infty$ and as m remains in a compact set.

Proof of Corollary 7.4. *(1)* We suppose that $\ell(\theta_i(m,\alpha),[K_i'])$ is increasing from $\alpha = k$, where k is an integer. We then take $\gamma_i = \rho^k([K_i'])$ and we apply Remark 2 above.

(2) This is a general property of families of functions of a real variable that are strictly convex and take a minimum, and that, in the compact–open topology, depend continuously on a parameter. Let $f_\lambda(x)$ be such a family, and let $m(\lambda)$ be the point that realizes the minimum. Then $m(\lambda)$ is a continuous function. Indeed, ϵ being given, if λ is sufficiently close to λ_0, we have

$$f_\lambda(m(\lambda_0)) < \inf\{f_\lambda(m(\lambda_0) - \epsilon), f_\lambda(m(\lambda_0) + \epsilon)\};$$

thus $m(\lambda)$ belongs to the open interval $(m(\lambda_0) - \epsilon, m(\lambda_0) + \epsilon)$.

Now, let $x_0 > m(\lambda_0)$ and let K lie between $f_{\lambda_0}(m(\lambda_0))$ and $f_{\lambda_0}(x_0)$. Then if λ is sufficiently close to λ_0, one has $f_\lambda(x_0) > K$ and f_λ is strictly increasing on $[x_0, \infty)$; thus $f_\lambda([x_0, \infty)) \subset (K, \infty)$. $\qquad\square$

LEMMA 7.5. *Let γ be a geodesic in the hyperbolic plane and let τ be an isometry leaving γ invariant. Let x be a point of γ and y a point not on γ; then*

$$d(x, \tau x) < d(y, \tau y),$$

where d denotes hyperbolic distance.

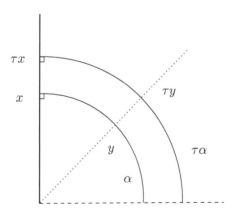

Figure 7.1.

Proof. We can take x to be the foot of the perpendicular α from y onto γ. Then γ is the unique common perpendicular to α and $\tau\alpha$. This gives the inequality. See Figure 7.1. $\qquad\square$

LEMMA 7.6. *Let γ_1 and γ_2 be two geodesics in the hyperbolic plane that do not intersect. Then the function $d(x,y)$, $x \in \gamma_1$, $y \in \gamma_2$, is strictly convex.*

Proof. Let x, x' and y, y' be pairs of points on γ_1 and γ_2, respectively (see Figure 7.2). Without loss of generality we suppose that $x \neq x'$. Let i be the midpoint of the arc $\overline{xx'}$, j that of $\overline{yy'}$, and δ the geodesic segment $\overline{ij}$. Denote by σ_i and σ_j the reflections through the points i and j. The product $\sigma_j\sigma_i$ is an isometry that leaves δ invariant. Let $z = \sigma_j\sigma_i(x)$, $z' = \sigma_j\sigma_i(x')$, and $k = \sigma_j\sigma_i(i)$. Then σ_j takes x to z' and y to y'. Therefore

$$d(x,y) = d(y', z').$$

Also, by the triangle inequality, we have

$$d(x', z') \leq d(x', y') + d(y', z').$$

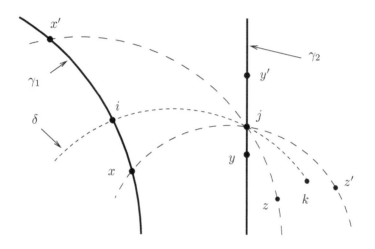

Figure 7.2.

By Lemma 7.5, we have

$$2d(i,j) = d(i,k) < d(x', z').$$

(Note that since γ_1 does not intersect γ_2, the point x' is not on δ.)

Finally, we obtain the inequality of convexity:

$$2d(i,j) < d(x,y) + d(x', y').$$

$\square$

Proof of Proposition 7.3. Identify the metric universal cover of the surface with $\mathbb{H}^2$. There exists an element τ of $\pi_1(M, *)$ that acts as an isometry of $\mathbb{H}^2$, leaving invariant a geodesic δ that lifts the geodesic K'_i. Let x_0 be a point of δ that projects to a point of $K_i' \cap K_i$. Denote by $\widetilde{K}^1$ the lift of K_i that passes through x_0 and by $\widetilde{K}^3$ the lift that passes through τx_0. The segment $(x_0, \tau x_0)$ intersects exactly one other lift $\widetilde{K}^2$ of K_i in a point y_0. Figure 7.3 shows the orientations of these three lifts.

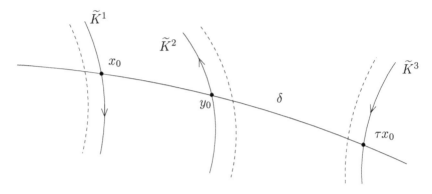

Figure 7.3.

If we twist the metric by an "angle α" in the collars indicated in the figure, the lift of the $\theta_i(m, \alpha)$-geodesic of $[K_i']$ intersects $\widetilde{K}^1$ in a point x_α and $\widetilde{K}^2$ in y_α; this is a geodesic from x_α to y_α in the metric of the hyperbolic plane. On the other hand, the part of this lift from y_α to τx_α has length given by the hyperbolic distance $d(y_\alpha + \alpha, \tau x_\alpha + \alpha)$; in this formula $+$ denotes translation along the geodesics $\widetilde{K}^2$ and $\widetilde{K}^3$. Finally, we have

$$\ell(\theta_i(m, \alpha), [K_i']) = \inf_{x \in \widetilde{K}^1, y \in \widetilde{K}^2} (d(x, y) + d(y + \alpha, \tau x + \alpha)).$$

We are going to show that $f(x, y, \alpha) = d(x, y) + d(y + \alpha, \tau x + \alpha)$ is a proper and strictly convex function. To do this, we use the fact that $d(x, y)$ is proper; this follows from the fact that the geodesics on which the points are moved have a common perpendicular (at a finite distance) and the fact that d is strictly convex (Lemma 7.6).

We now show the properness of f. Let $(x_n, y_n, \alpha_n) \to \infty$. If $(x_n, y_n) \to \infty$, then $d(x_n, y_n) \to \infty$, hence $f(x_n, y_n, \alpha_n) \to \infty$. If (x_n, y_n) stays in a compact set, then $\alpha_n \to \infty$ and $(y_n + \alpha_n, \tau x_n + \alpha_n)$ tends to ∞, hence $d(y_n + \alpha_n, \tau x_n + \alpha_n)$ tends to ∞.

One verifies immediately that f is strictly convex.

For α fixed, the function $f(x, y, \alpha)$ has a minimum $h(\alpha)$, by the properness of f. The convexity of f implies that h is also convex; since $h(\alpha)$ is a value attained by $f(x, y, \alpha)$, we see that h is strictly convex.

The function f has an absolute minimum (f is proper and bounded below); it is the minimum of h.   □

PROPOSITION 7.7. *The map* $L\colon \mathcal{T} \to (\mathbb{R}_+^*)^{3g-3}$ *is a principal fibration for the group* $\mathbb{R}^{3g-3}$ *acting by* θ.

COROLLARY 7.8. *The Teichmüller space of a closed surface of genus g is homeomorphic to* $\mathbb{R}^{6g-6}$.

Proof. The important point is to show that there exist local sections for L. We know from Theorem 3.5 that, for the model pair of pants P^2, the map

$$\mathcal{H}(P^2) \to (\mathbb{R}_+^*)^3$$

which, to a metric adapted to the boundary, associates the three lengths of the boundary, admits local sections on the level of metrics.

We know that to glue together two hyperbolic metrics along a geodesic, it is enough to specify an isometry of the geodesic along which we glue. Now, if one has a metric on P^2 and if one considers a curve C of the boundary, there is a unique simple geodesic arc that meets C in its two endpoints (a *pants seam*[1]). By Proposition 7.1, its origin (which one distinguishes from the other endpoint by an orientation chosen once and for all) varies continuously with the metric.

The desired local section is now obtained as follows. Above the $3g-3$ lengths one chooses a metric with the following property: if K_j is adjacent to R_{i_1} and R_{i_2}, the two origins on K_j of the pants seams of the two pairs of pants coincide. By imposing this condition, we obtain a continuous local section.

Let D be a ball of $(\mathbb{R}_+^*)^{3g-3}$ over which L admits a section σ. Define a map $T\colon D \times \mathbb{R}^{3g-3} \to \mathcal{T}$ by

$$T(x, \alpha_1, \ldots, \alpha_{3g-3}) = \theta(\sigma(x), \alpha_1, \ldots, \alpha_{3g-3}).$$

It remains to show that T is a homeomorphism onto its image. Since $\mathcal{T}$ is second countable, it is enough to show that T is injective and proper.

If two metrics differ from one another by a twist, they are distinguished by the length of a geodesic (Corollary 7.4); this proves injectivity.

For simplicity, denote $(\alpha_1, \ldots, \alpha_{3g-3})$ by α. Let (x^n, α^n) be a sequence tending to infinity in $D \times \mathbb{R}^{3g-3}$. The second part of the same corollary gives that the image under T of this sequence cannot be a compact set in Teichmüller space. Hence T is proper. □

THEOREM 7.9. *The map* $\ell_*\colon \mathcal{T} \to \mathbb{R}_+^{\mathcal{S}}$ *is a proper map that is a homeomorphism onto its image.*

Actually, we are going to prove a stronger statement (Proposition 7.11 below), relative to the system of curves K_i, K_i', K_i'' described before Proposition 7.3. First, we need a lemma.

LEMMA 7.10. *For any sequence of hyperbolic metrics where the length of some K_i tends to 0, the lengths of K_i' and K_i'' tend to infinity.*

---

[1]Compare with the terminology for measured foliations (Exposé 6).

Proof. The lemma follows immediately from the inequality

$$\cosh(l(m, [K_i'])) \sinh(l(m, [K_i])) \geq 1. \tag{7.1}$$

One can establish the inequality (7.1) from Formula 7.18.2 in [Bea83]. We will show how it follows from Lemma D.4 (below).

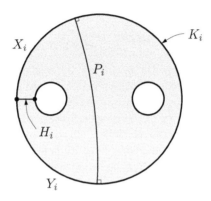

Figure 7.4.

With the notation of Figure 7.4, the length of K_i' is larger than that of the "bridge" P_i and the length of K_i is larger than the length of X_i and that of Y_i. If we cut along P_i and H_i, we obtain a right-angled hexagon which admits X_i, P_i, Y_i as consecutive sides. The formula from Lemma D.4 then shows that

$$\cosh(l(m, P_i)) \sinh(l(m, X_i)) \sinh(l(m, Y_i))$$
$$\geq 1 + \cosh(l(m, X_i)) \cosh(l(m, Y_i)),$$

from which it follows that

$$\cosh(l(m, P_i)) \sinh(l(m, X_i)) \geq 1.$$

The inequality (7.1) follows. It is clear from this proof that this formula is not optimal. We remark that this is a particular case of the "collar lemma" or the Margulis inequality. □

PROPOSITION 7.11. *The map* $\Lambda \colon \mathcal{T} \to \mathbb{R}_+^{9g-9}$ *that to* $m \in \mathcal{T}$ *associates the tuple*

$$(\ell(m, [K_i]), \ell(m, [K_i']), \ell(m, [K_i'']))$$

is injective and proper (hence a homeomorphism onto its image).

Proof. We choose a section s of the fibration L; that is, we write every $m \in \mathcal{T}$ in the form

$$m = \theta(s(x), \alpha)$$

where $\alpha = (\alpha_1, \ldots, \alpha_{3g-3}) \in \mathbb{R}^{3g-3}$ is a "multi-angle" and where $x \in (\mathbb{R}_+^*)^{3g-3}$ is the tuple of lengths of the curves K_i.

The variable x being fixed, the function $\ell(m, [K_i'])$ is a strictly convex and proper function $g_i(\alpha_i)$ of the ith component of α. Moreover, $\ell(m, [K_i'']) = g_i(\alpha_i + 1)$.

We have the following fact:

> If $g: \mathbb{R} \to \mathbb{R}$ is a strictly convex proper function, then $t \mapsto (g(t), g(t+1))$ defines a proper immersion of $\mathbb{R}$ into $\mathbb{R}^2$.

Thus, the $(6g - 6)$-tuple $(\ell(m, [K_i']), \ell(m, [K_i'']))$ is an injective proper function of the multi-angle α. From this, it follows that Λ is injective.

To show that Λ is proper, we consider a sequence (x_n, α_n) tending to infinity. If x_n tends to ∞, then either the length of some K_i tends to 0, or the length of some K_i tends to infinity; it follows from Lemma 7.10 that in either case $\Lambda(x_n, \alpha_n)$ tends to ∞. If, on the other hand, the x_n remain in a compact set, then, by Corollary 7.4, the length of one of the curves K_i' tends to ∞. □

We complete this exposé with the following proposition, following a proof indicated by S. Kerckhoff. Recall that π denotes the projection $\mathbb{R}^{\mathcal{S}} - \{0\} \to P(\mathbb{R}^{\mathcal{S}})$.

PROPOSITION 7.12. *The composite map* $\pi \circ \ell_*: \mathcal{T} \to P(\mathbb{R}_+^{\mathcal{S}})$ *is an injection.*

To prove the proposition, we need two lemmas from hyperbolic geometry. We use the upper half-plane model for $\mathbb{H}^2$, $\{x + iy \mid y > 0\}$, with the metric $ds^2 = (dx^2 + dy^2)/y^2$. The group of isometries is $\mathrm{PSL}(2, \mathbb{R}) = \mathrm{SL}(2, \mathbb{R})/\{\pm \mathrm{Id}\}$, where the action of $A = \left(\begin{smallmatrix} a & b \\ c & d \end{smallmatrix}\right)$ is given by $z \mapsto (az + b)/(cz + d)$.

If A is a hyperbolic element (i.e., it leaves invariant a geodesic), we define its *displacement* as
$$\ell(A) = \inf_{z \in \mathbb{H}^2} d(z, A \cdot z).$$

The minimum is attained on the invariant geodesic.

LEMMA 7.13. *If* $A \in \mathrm{SL}(2, \mathbb{R})$ *is hyperbolic, we have*

$$\mathrm{Tr}(A) = 2 \cosh\left(\frac{\ell(A)}{2}\right).$$

Proof. By conjugating within $\mathrm{SL}(2, \mathbb{R})$, we reduce to the case where the invariant geodesic is the positive y-axis. We then have

$$A = \left(\begin{array}{cc} \rho & 0 \\ 0 & \rho^{-1} \end{array}\right), \qquad \rho > 0.$$

As a consequence, $A \cdot i = \rho^2 i$. Thus, we have

$$\ell(A) = d(i, \rho^2 i) = \int_1^{\rho^2} \frac{dt}{t} = 2 \log(\rho)$$

and

$$\operatorname{Tr}(A) = \rho + \rho^{-1} = 2\cosh\left(\frac{\ell(A)}{2}\right).$$

□

LEMMA 7.14. *Let $A, B \in \mathrm{SL}(2, \mathbb{R})$. We have*

$$\operatorname{Tr}(A) \cdot \operatorname{Tr}(B) = \operatorname{Tr}(AB) + \operatorname{Tr}(A^{-1}B).$$

The proof is a direct calculation.

To prove Proposition 7.12, we need the following technical lemma.

LEMMA 7.15. *Let α, β, γ, and δ be four nonnegative numbers and let k be a positive number different from 1. If*

$$\begin{aligned}
\cosh\alpha + \cosh\beta &= \cosh\gamma + \cosh\delta \quad \text{and} \\
\cosh k\alpha + \cosh k\beta &= \cosh k\gamma + \cosh k\delta,
\end{aligned}$$

then $\{\alpha, \beta\} = \{\gamma, \delta\}$.

Proof. One may restrict to the case $k > 1$. The reader may check that the function $\cosh(k\cosh^{-1}x)$ is a strictly convex function of x. Now, if c is a common value of the first equality and if one sets $x = \cosh\alpha$ and $y = \cosh\gamma$, the second relation is

$$\begin{aligned}
&\cosh(k\cosh^{-1}x) + \cosh(k\cosh^{-1}(c-x)) \\
&= \cosh(k\cosh^{-1}y) + \cosh(k\cosh^{-1}(c-y)).
\end{aligned}$$

We may suppose that $y \le x \le c - x \le c - y$. If $y < x$, then by the strict convexity, the left side will be strictly less than the right. □

Proof of Proposition 7.12. Consider on the surface M two simple oriented curves γ_1 and γ_2 that intersect transversely at the basepoint. The homotopy classes of based loops $\gamma_1 * \gamma_2$ and $\gamma_1^{-1} * \gamma_2$ can both be represented by simple curves γ_3 and γ_4. If M is given a metric m of curvature -1, these elements of the fundamental group correspond to hyperbolic isometries of $\mathbb{H}^2$ for which the displacement is $\ell_i = \ell(m, [\gamma_i])$. The preceding lemmas thus give the formula

$$2\cosh\left(\frac{\ell_1}{2}\right)\cosh\left(\frac{\ell_2}{2}\right) = \cosh\left(\frac{\ell_3}{2}\right) + \cosh\left(\frac{\ell_4}{2}\right)$$

or

$$\begin{aligned}
\cosh\left(\frac{\ell_1 + \ell_2}{2}\right) &+ \cosh\left(\frac{\ell_1 - \ell_2}{2}\right) \\
&= \cosh\left(\frac{\ell_3}{2}\right) + \cosh\left(\frac{\ell_4}{2}\right).
\end{aligned} \qquad (7.2)$$

For purposes of contradiction, we make the following hypothesis:

(H) Suppose that there is another metric of curvature -1 for which the lengths of all closed geodesics are multiplied by $k \neq 1$.

For such a metric, the equality (7.2) becomes

$$\cosh\left(k\frac{\ell_1 + \ell_2}{2}\right) + \cosh\left(k\frac{\ell_1 - \ell_2}{2}\right) \qquad (7.3)$$
$$= \cosh\left(k\frac{\ell_3}{2}\right) + \cosh\left(k\frac{\ell_4}{2}\right).$$

Applying Lemma 7.15, (7.2) and (7.3) give

$$\{\ell_1 + \ell_2, \ell_1 - \ell_2\} = \{\ell_3, \ell_4\}.$$

Up to change of notation, we can say that

$$\ell_3 = \ell_1 + \ell_2.$$

Since the angle between γ_1 and γ_2 is nonzero, it is not possible for $\ell_1 + \ell_2$ to be a shorter distance; hence, the above equality cannot be true, and the hypothesis (H) is absurd. □

Exposé Eight

The Thurston Compactification of Teichmüller Space

by Albert Fathi and François Laudenbach

In Exposés 6 and 7, we showed that the Teichmüller space $\mathcal{T}$ and the space of Whitehead classes of measured foliations $\mathcal{MF}$ both embed into the space of functionals $\mathbb{R}_+^{\mathcal{S}}$. In this exposé, we identify these spaces with their images in the functional space. For any functional $f \in \mathbb{R}_+^{\mathcal{S}}$, we denote by $i(f, \alpha)$ the value of the functional on $\alpha \in \mathcal{S}$.

Recall that $\pi : \mathbb{R}_+^{\mathcal{S}} - \{0\} \to P(\mathbb{R}_+^{\mathcal{S}})$ denotes projection onto the space of rays and that $\mathcal{PMF}$ is the image of $\mathcal{MF}$; moreover, $\mathcal{MF} = \pi^{-1}(\mathcal{PMF})$. We will construct a topology on the union of $\mathcal{T}$ and $\mathcal{PMF}$. We prove that the resulting space is a manifold with boundary. Since the interior is homeomorphic to an open ball and the boundary is homeomorphic to a sphere, the manifold with boundary is homeomorphic to a closed ball.

The key is in the inequalities of the "fundamental lemma" below. The proof of this lemma relies on length estimates from hyperbolic geometry, gathered in the appendix to this exposé.

8.1 PRELIMINARIES

PROPOSITION 8.1. *In $\mathbb{R}_+^{\mathcal{S}}$, the spaces $\mathcal{T}$ and $\mathcal{MF}$ are disjoint.*

Proof. If f belongs to $\mathcal{T}$, then, since the surface is compact, the set of numbers $i(f, \alpha)$, $\alpha \in \mathcal{S}$, is bounded below by a strictly positive constant. We are going to prove that, for $f \in \mathcal{MF}$, the closure of the set of numbers $i(f, \alpha)$ contains zero.

Let $(\mathcal{F}, \mu)$ be a measured foliation representing f. Given any $\epsilon > 0$, we can choose an arc γ that is transverse to $\mathcal{F}$ and has measure $\mu(\gamma) \leq \epsilon$. By Poincaré recurrence (Theorem 5.2) almost every leaf departing from a point of γ returns to γ. We thus obtain a simple closed curve γ' formed by an arc of γ and an arc carried by a leaf of $\mathcal{F}$. If α is the isotopy class of γ', we have

$$i(f, \alpha) \leq \mu(\gamma') \leq \mu(\gamma) \leq \epsilon. \qquad \square$$

Construction of a projection $q : \mathcal{T} \to \mathcal{MF}$. The projection we construct will give the charts for the manifold-with-boundary structure. It depends on the choice of a family $\mathcal{K} = \{K_1, \ldots, K_k\}$ of mutually disjoint simple closed curves

118

cutting the surface into embedded pairs of pants $R_1, \ldots, R_{k'}$. If the surface is closed and of genus $g \geq 2$, then $k = 3g - 3$ and $k' = 2g - 2$.

Let $m \in \mathcal{T}$. We represent m by a metric $\overline{m}$, of curvature -1, for which the curves K_j are geodesics. The foliation that will represent $q(m)$ will be transverse to each K_j. For each j, we specify

$$i(q(m), K_j) = i(m, K_j).$$

Let R be one of the pairs of pants. Say that $\partial R = K_1 \cup K_2 \cup K_3$, and set $2m_j = i(m, K_j)$. Let $g_{jj'}$ be the simple $\overline{m}$-geodesic in R orthogonal to K_j and to $K_{j'}$.

Case 1: (m_1, m_2, m_3) satisfies the triangle inequality. Let T_{12} be the (closed) geodesic tube of points in R at a distance from g_{12} at most $(m_1 + m_2 - m_3)/2$. This tube is foliated by equidistant lines and the distance between two leaves gives the transverse measure. We consider in the same way the foliated tubes T_{23} and T_{13}. Each pair of tubes has exactly two points of intersection, which are on the boundary; for example,

$$T_{12} \cap T_{13} = T_{12} \cap T_{13} \cap K_1.$$

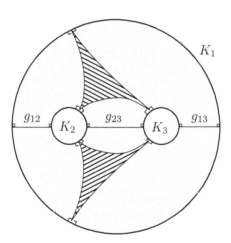

Figure 8.1. Case 1

This is because K_1 is the unique common perpendicular to g_{12} and g_{13}. Further, by adding the two thicknesses, we see that K_1 is totally covered; the same for K_2 and K_3. We obtain in this way a *partial measured foliation* of R (Figure 8.1). We then obtain a true measured foliation by collapsing each nonfoliated triangle to a tripod. Actually, for what follows, we are interested in keeping the partial foliation, in which the measure is directly given by the metric.

Up to renumbering, there is one other case.

Case 2: $m_1 > m_2 + m_3$.   Here T_{12} is the tube of radius m_2 and T_{13} is the tube of radius m_3. The set of points of K_1 that are in the complement of the interiors of T_{12} and T_{13} consists of two arcs A and A'. There is an isometric involution of R that interchanges A and A' and has $g_{12} \cup g_{13} \cup g_{23}$ as the locus of fixed points (Lemma 3.7). Let T_{11} be the union of lines of equal distance to the geodesic g_{11}, emanating from A (Warning! g_{11} might not lie in T_{11}). We see that $T_{11} \cap K_1 = A \cup A'$. These three tubes give a partial foliation that looks like the one in Figure 8.2.

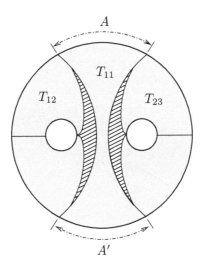

Figure 8.2. Case 2

We remark that, in the two cases, the leaves are perpendicular to the curves of the boundary. When we reglue the pairs of pants, we obtain a partial measured foliation $\mathcal{F}_{\overline{m}}$, which represents $q(m)$. The leaves are only C^1 at the junction of two pairs of pants, but this is not important.

PROPOSITION 8.2. *The map q is a homeomorphism of $\mathcal{T}$ onto the open set $\mathcal{U}(\mathcal{K})$ of $\mathcal{MF}$ consisting of the functionals taking nonzero values on each component of $\mathcal{K}$.*

Proof. We first construct an inverse map q^{-1} as follows. An element of $\mathcal{U}(\mathcal{K})$ is represented by a measured foliation $(\mathcal{F}, \mu)$ that is transverse to the curves of $\mathcal{K}$.

In the pair of pants R (notation from above), we construct a metric $\overline{m}$ of curvature -1 with the following properties:

(i) $\overline{m}|_{K_j} = \mu|_{K_j}$, for $j = 1, 2, 3$.

(ii) Denoting $2m_j = \mu(K_j)$, if $(m_1, m_2, m_3) \in (\nabla \le)$, the smooth leaf that goes from K_1 to K_2 (resp. to K_3) and whose μ-distance to the singularities is $(m_1 + m_2 - m_3)/2$ (resp. $(m_1 + m_3 - m_2)/2$) is declared to be a geodesic of $\overline{m}$ orthogonal to the boundary.

(iii) If $m_1 > m_2 + m_3$, the smooth leaf that goes from K_1 to K_2 (resp. to K_3) and whose μ-distance to a singularity is m_2 (resp. m_3), is declared to be a geodesic of $\overline{m}$ orthogonal to the boundary.

By the classification of hyperbolic metrics of pairs of pants (Theorem 3.5), if two metrics satisfy the conditions above, then they are conjugate by a diffeomorphism isotopic to the identity, by an isotopy that is constant on the boundary. Therefore, when we glue all the pairs of pants, we obtain a hyperbolic metric that is well-defined up to isotopy. By the classification of measured foliations on pairs of pants (Proposition 6.7), we see that the map constructed in this way is the inverse of q.

For the continuity of q and q^{-1}, we proceed in the following manner. We utilize the parametrization $\{m_j, s_j, t_j\}$ of $\mathcal{MF}$ (see Exposé 6). The projection $\{m_j, s_j, t_j\} \to \{m_j\}$ restricted to $\mathcal{U}(\mathcal{K})$ is a principal bundle for which the structure group is the group of twists along $\mathcal{K}$. Indeed, we have an obvious section $\sigma(\{m_j\}) = \{m_j, 0, m_j\}$; moreover, if we act by a twist α_j along $\mathcal{K}_j$ on this section, the pair (s_j, t_j) parametrizing the twisted foliation is given by semilinear formulas (exercise). This establishes, for each m_j, a homeomorphism of $\mathbb{R}$ onto the set of (s_j, t_j) such that (m_j, s_j, t_j) belongs to $\partial(\nabla \le)$. Since $\mathcal{U}(\mathcal{K})$ is a manifold, these arguments suffice to ensure the structure of a principal bundle.

We also recall that $\mathcal{T}$ is fibered over the space of lengths of the components of $\mathcal{K}$ (Proposition 7.7). By construction, the map q is equivariant with respect to these two principal bundle structures, and it extends the identity map of their common base.

The continuity of q is equivalent to that of q^{-1}, since the source and the target are manifolds. For the continuity of q^{-1}, by the above, it suffices to verify this on the section σ. However, over the closed set $(\nabla \le) \cup \{m_1 \ge m_2 + m_3\}$, the section σ lifts to a section $\tilde{\sigma}$ with values in the space of foliations on the pair of pants R, where the middle leaves (specified in *(ii)* and *(iii)*) are fixed. Starting from this, we can construct $\overline{m}$ continuously in R, by applying Theorem 3.5. We do the same for all the pairs of pants. $\qquad\square$

8.2   THE FUNDAMENTAL LEMMA

LEMMA 8.3 (The fundamental lemma). *Let $\varepsilon > 0$ and let $V(\mathcal{K}, \varepsilon)$ be the open subset of $\mathcal{T}$ defined by the metrics for which each component of $\mathcal{K}$ is a* *geodesic of length > ε. For each α ∈ S, there exists a constant C such that, for all m ∈ V(K, ε), we have*

$$i(q(m), \alpha) \leq i(m, \alpha) \leq i(q(m), \alpha) + C.$$

Proof. First we show that $i(q(m), \alpha) \leq i(m, \alpha)$. If $\overline{m}$ is a metric representing m, the transverse measure of the foliation $\mathcal{F}_{\overline{m}}$, constructed as a representative of $q(m)$, is given by the metric $\overline{m}$ on geodesics orthogonal to the leaves. Therefore, the $\overline{m}$-length of an arc is at least as big as the $\mathcal{F}_{\overline{m}}$-measure. Further, by the definition of the functional, the $\mathcal{F}_{\overline{m}}$-measure of a closed curve of the class α bounds $i(q(m), \alpha)$ from above, which proves the first inequality.

Next we show that $i(m, \alpha) \leq i(q(m), \alpha) + C$. It suffices to prove it on the dense subset of $V(K, \varepsilon)$ consisting of those m for which the foliation $\mathcal{F}_{\overline{m}}$ has simple (tripod) singularities without connections between these singularities; such classes of metrics are called *generic*.

By Proposition 5.9, if m is generic, then α can be represented by a simple curve α' transverse to the foliation $\mathcal{F}_{\overline{m}}$. Its measure $\mathcal{F}_{\overline{m}}(\alpha')$ is $i(q(m), \alpha)$. We can, moreover, choose α' so that for all j we have

$$\text{card}(K_j \cap \alpha') = i([K_j], \alpha).$$

In fact, if this is not already the case, there is a disk whose boundary is formed by an arc of α' and an arc of K_j. Since each of these is transverse to the foliation, the disk is foliated as in Figure 8.3 and the assertion is clear.

Figure 8.3.

Now, two curves that are isotopic and in minimal position with K are isotopic by an isotopy leaving K invariant (Proposition 3.13). Therefore α' is cut by K into n arcs:

$$\alpha' = \alpha'_1 \cup \alpha'_2 \cup \cdots \cup \alpha'_n,$$

where n depends only on α, each α'_j is an essential arc of one of the pairs of pants of the decomposition, and each α'_j is transverse to the foliation. The inequality is therefore a consequence of Lemma 8.4. $\qquad\qquad\square$

LEMMA 8.4. *Let $\varepsilon > 0$. There exists a constant C' with the following property. For each hyperbolic metric $\overline{m}$ on the pair of pants P^2 where each component of the boundary is a geodesic of length $\geq \varepsilon$, and for each simple arc β of P^2 going from boundary to boundary transverse to the foliation $\mathcal{F}_{\overline{m}}$, there exists an arc γ, homotopic to β with endpoints fixed, such that the $\overline{m}$-length of γ is less than or equal to $\mathcal{F}_{\overline{m}}(\beta) + C'$.*

Proof. We consider separately each type of foliation (Figures 8.1 and 8.2) and we take the bigger of the constants. We first do the argument for the foliation satisfying the triangle inequality.

We replace β by an immersed arc β' with the same endpoints, by applying the two processes shown in Figure 8.4.

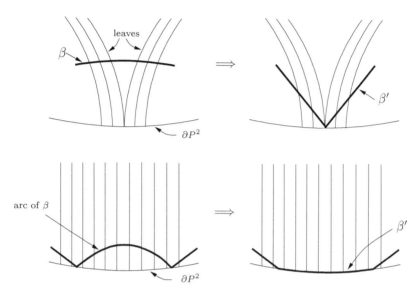

Figure 8.4. Replacing β by β'

We remark that β' is transverse to $\mathcal{F}_{\overline{m}}$, with

$$\mathcal{F}_{\overline{m}}(\beta') = \mathcal{F}_{\overline{m}}(\beta),$$

and that β' is close to a simple arc. By construction, β' is formed from an arc of the boundary and from "diagonals" in the foliated rectangles (the first and last endpoints of β' might not lie on vertices of rectangles, so we need to extend the usual definition of a diagonal of a rectangle in these cases). We deduce from the topology that β' contains at most three diagonals (each covered one time). For example, if δ_1 is the first diagonal found along β' (see Figure 8.5), then δ_2 is necessarily the second, and δ_3 the third. Upon leaving δ_3, the arc β' travels along the boundary in such a way as to make it impossible to traverse any of the diagonals again.

We replace each diagonal by an arc of a leaf and an arc of the boundary. In this way we obtain an arc β'' with the same $\mathcal{F}_{\overline{m}}$-measure and containing at most three leaves. Finally we form γ by replacing the leaves by the geodesics with the same endpoints. The length of γ is the sum of the lengths of the geodesics and the lengths of the arcs along the boundary. The latter term has

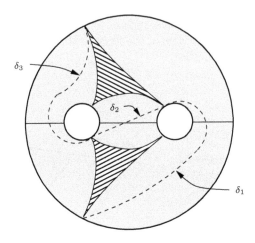

Figure 8.5.

value $\mathcal{F}_{\overline{m}}(\beta'') = \mathcal{F}_{\overline{m}}(\beta)$, and the contribution of the first term is bounded, by Proposition D.5 in Appendix D.

If $\mathcal{F}_{\overline{m}}$ is the foliation of Figure 8.2, then β' contains at most three diagonals (Figure 8.6); to bound from above the length of a geodesic joining the endpoints of a leaf of the tube T_{11}, one must use Corollary D.6 in Appendix D. $\square$

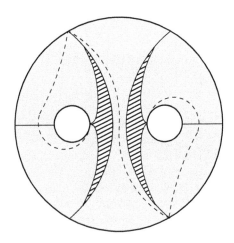

Figure 8.6.

COROLLARY 8.5. *Let x_n be a sequence in $V(\mathcal{K}, \varepsilon)$ tending to infinity in $\mathcal{T}$. Then $\pi(x_n)$ converges if and only if $\pi \circ q(x_n)$ converges, and in this case the* *two sequences have the same limit.*

Proof. Say that $\pi(x_n)$ converges. That is to say that there exists a sequence of scalars $\lambda_n > 0$ such that the sequence $\lambda_n x_n$ converges. Since the topology of $\mathcal{T}$ is determined by a finite number of curves $\gamma_1, \ldots, \gamma_k$, we have that

$$\sum_j i(x_n, \gamma_j) \to \infty$$

and $\sum \lambda_n i(x_n, \gamma_j)$ converges. Therefore, $\lambda_n \to 0$.
By Lemma 8.3, for all $\alpha \in \mathcal{S}$, we have

$$|i(\lambda_n \, x_n, \alpha) - i(\lambda_n \, q(x_n), \alpha)| \to 0,$$

so $\pi \circ q(x_n)$ converges to the same limit as $\pi(x_n)$. The converse is analogous. □

8.3 THE MANIFOLD $\overline{\mathcal{T}}$

Topology. On the disjoint union $\mathcal{T} \cup \mathcal{PMF}$, we take as a basis the open sets of $\mathcal{T}$ (open sets of type 1), and the sets of the form $(\mathcal{T} \cap \pi^{-1}(U)) \cup (\mathcal{PMF} \cap U)$, where U is an open set of the projective space (open sets of type 2). As $\pi^{-1}(U) \cap \mathcal{T}$ is an open set of $\mathcal{T}$, the intersection of an open set of type 1 and an open set of type 2 is an open set of type 1. We then easily verify the axioms of a topology. This topological space is denoted $\overline{\mathcal{T}}$.

The space $\mathcal{T}$ is equipped with a continuous map to the projective space which is an injection; in fact, π injects $\mathcal{T}$ (Proposition 7.12) and $\pi(\mathcal{T})$ avoids $\mathcal{PMF}$ by Proposition 8.1. In particular, $\overline{\mathcal{T}}$ is a separable space. The topology of $\overline{\mathcal{T}}$ is second countable.

Map of the neighborhood of a foliation. Let $f \in \mathcal{PMF}$. By Lemma 6.16, there exists a decomposition of the surface into pairs of pants along a system $\mathcal{K}$ of curves $K_1, \ldots, K_k$ such that $i(\overline{f}, [K_j]) \neq 0$ for all j, where $\overline{f}$ denotes some lift of f to $\mathbb{R}_+^{\mathcal{S}}$ (being nonzero is a projective property). Let $\{K_j', K_j''\}$ be a system of curves that, together with the $\{K_j\}$, parametrize $\mathcal{T}$ (Proposition 7.11).

Let $\varepsilon > 0$ be arbitrary. We consider the open set $V(\mathcal{K}, \varepsilon)$ of $\mathcal{T}$ (see Lemma 8.3) and the open set W of $\mathcal{PMF}$ of the "projective" functionals that are nonzero on the components of $\mathcal{K}$; we have $\pi^{-1}(W) = \mathcal{U}(\mathcal{K})$ (see Proposition 8.2) and $\pi \circ q(V(\mathcal{K}, \varepsilon)) = W$. We define

$$\phi : W \cup V(\mathcal{K}, \varepsilon) \longrightarrow W \times [0, 1]$$

by

$$\phi(x) = \begin{cases} (x, 0) & x \in W, \\ \left(\pi \circ q(x), e^{-\Sigma\{i(q(x), K_j) + i(q(x), K_j') + i(q(x), K_j'')\}}\right) & x \in V(\mathcal{K}, \varepsilon). \end{cases}$$

LEMMA 8.6. *We have:*
(i) $W \cup V(\mathcal{K}, \varepsilon)$ *is an open set of* $\overline{\mathcal{T}}$.
(ii) ϕ *is a homeomorphism onto an open subset of* $W \times [0, 1]$.

Proof. *(i)* Let $x \in W$. Suppose that the designated set is not a neighborhood of x in $\overline{\mathcal{T}}$. Then there exists a sequence x_n in $\mathcal{T}$, $x_n \notin V(\mathcal{K}, \varepsilon)$, such that $\pi(x_n)$ tends to x. Up to change of indices and extraction of a subsequence, one can say that $i(x_n, K_1) \leq \varepsilon$.

By Proposition 8.1, the sequence x_n does not have a subsequence converging to a point of $\mathcal{T}$. Therefore x_n tends to infinity. Moreover, there exists a sequence of scalars $\lambda_n > 0$ such that $\lambda_n x_n$ converges to a measured foliation f in the projective class of x. We deduce that $\lambda_n \to 0$. But then $i(f, K_1) = 0$, which contradicts the assumption that $x \in W$.

(ii) Step 1: The map ϕ is continuous in $x \in W$. Suppose $x_n \in V(\mathcal{K}, \varepsilon)$ converges to $x \in W$. By the proof of part *(i)* of the lemma, we have that x_n tends to infinity in $\mathcal{T}$. As x_n belongs to $V(\mathcal{K}, \varepsilon)$, Corollary 8.5 implies that the first component of $\phi(x_n)$ converges to x. For the same reason,

$$\sum_j \left(i(q(x), K_j) + i(q(x), K_j') + i(q(x), K_j'') \right)$$

tends to ∞. Therefore the second component of $\phi(x_n)$ tends to 0.

Step 2: The map ϕ is injective. A priori, a failure of injectivity can only come from two elements x and y of $\mathcal{T}$. If $q(x) = \lambda q(y)$, the equality in the second coordinate implies $\lambda = 1$ (see Figure 8.7). But q is injective.

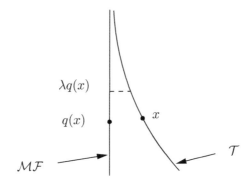

Figure 8.7.

We remark that if $\phi(x) = (z, t)$, then $\phi(q^{-1}(\lambda q(x))) = (z, t^\lambda)$.

Step 3: There exists a continuous section $\sigma : W \to \mathcal{MF}$ of π. Up to multiplication by a scalar, one can take it to have values in $q(V(\mathcal{K}, \varepsilon))$. The manifold $\phi \circ q^{-1} \circ \sigma(W)$ is the graph in $W \times [0, 1)$ of a strictly positive function defined on W. The neighborhood of $W \times \{0\}$ bounded by this graph is surely in the image of ϕ by the preceding remark.

Step 4: The inverse of ϕ is continuous on this neighborhood. We verify only that if $(z_n, t_n) \to (z, 0)$, then $\pi \circ \phi^{-1}(z_n, t_n)$ converges to z in the projective space. As t_n tends to 0, $q \circ \phi^{-1}(z_n, t_n)$ tends to infinity. By Lemma 8.3, $\phi^{-1}(z_n, t_n)$ tends to infinity in $\mathcal{T}$. We know that $z_n = \pi \circ q \circ \phi^{-1}(z_n, t_n)$ converges to z; therefore $\pi \circ \phi^{-1}(z_n, t_n)$ goes to the same limit by Corollary 8.5. $\square$

$\overline{\mathcal{T}}$ *is a ball.* We already know that $\mathcal{T}$ is a manifold, and we just saw that $\overline{\mathcal{T}}$ is a manifold with boundary, bounded by $\mathcal{PMF}$. In particular, $\overline{\mathcal{T}}$ is locally compact and, as the topology is second countable, it is paracompact. Thus, the boundary admits a collar neighborhood (this is a theorem of M. Brown, see [Rus73], Chapter 1, Theorem 17.4, page 40). As $\mathcal{PMF}$ is homeomorphic to a sphere, the interior boundary of the collar neighborhood is an embedded sphere in the interior of $\mathcal{T}$, hence in a Euclidean space. Then, by the Schöenflies theorem, generalized by Mazur and Brown ([Rus73], Chapter 1, Theorem 18.2, page 48), this sphere bounds a ball. Finally, $\overline{\mathcal{T}}$ is homeomorphic to a ball. In particular, it is a compact set. By Propositions 7.12 and 8.1, the projection $\pi : \mathbb{R}_+^\mathcal{S} - \{0\} \to P(\mathbb{R}_+^\mathcal{S})$ induces a continuous injection of $\overline{\mathcal{T}}$ into $P(\mathbb{R}_+^\mathcal{S})$, which is therefore a homeomorphism onto its image.

We have finally proven the following theorem.

THEOREM 8.7. *The space $\overline{\mathcal{T}} = \mathcal{PMF} \cup \mathcal{T}$, given the topology induced by $P(\mathbb{R}_+^\mathcal{S})$, is a compact manifold with boundary, homeomorphic to a ball, and bounded by $\mathcal{PMF}$.*

If the surface is closed and of genus $g \geq 2$, $\overline{\mathcal{T}}$ is homeomorphic to D^{6g-6}.

The group of isotopy classes of diffeomorphisms of the surface acts continuously on $\overline{\mathcal{T}}$, by the transposed action of the direct image action on $\mathcal{S}$.

Remark. The described action thus is the opposite of the one usually used on measured foliations, given by taking the direct images of the measures.

Appendix D

Estimates of Hyperbolic Distances

by Albert Fathi

We consider the Poincaré half-plane $\mathbb{H}^2 = \{z \in \mathbb{C} | z = x + iy, \, y > 0\}$, endowed with the metric $ds^2 = (dx^2 + dy^2)/y^2$. The distance between two points z and z' is denoted $d(z, z')$.

D.1   THE HYPERBOLIC DISTANCE FROM i TO A POINT z_0

LEMMA D.1. *The hyperbolic distance between the point i and an arbitrary point z_0 is given by the formula*

$$\cosh(d(i, z_0)) = \frac{|z_0 + i|^2 + |z_0 - i|^2}{|z_0 + i|^2 - |z_0 - i|^2}.$$

Proof. Let $f : \mathbb{H}^2 \to \mathbb{D}^2$ be the isomorphism

$$z \mapsto \frac{z - i}{z + i}.$$

Let g_{z_0} be the automorphism of $\mathbb{D}^2$ which is multiplication by $\overline{f(z_0)}/|f(z_0)|$.

We verify that the automorphism $f^{-1} \circ g_{z_0} \circ f$ of $\mathbb{H}^2$ fixes i and sends z_0 to the purely imaginary point

$$\frac{|z_0 + i| + |z_0 - i|}{|z_0 + i| - |z_0 - i|} i.$$

As we have the formula $d(i, iy) = |\log y|$, it follows that

$$e^{d(i,z_0)} = \frac{|z_0 + i| + |z_0 - i|}{|z_0 + i| - |z_0 - i|}.$$

$\square$

COROLLARY D.2. *If $z_0 = (ai + b)/(ci + d)$ with $ad - bc = 1$, we have*

$$\cosh(d(i, z_0)) = \frac{a^2 + b^2 + c^2 + d^2}{2}$$

Hyperbolic translations along the imaginary axis. The transformation

$$\begin{pmatrix} e^{k/2} & 0 \\ 0 & e^{-k/2} \end{pmatrix}$$

sends the geodesic iy to itself and the points on this line are displaced by a distance k.

Translation along the hyperbolic geodesic of complex numbers of modulus 1. If z has modulus 1 and if z' satisfies

$$\frac{z'+1}{z'-1} = e^k \frac{z+1}{z-1},$$

then z' has modulus 1. The transformation $z \mapsto z'$ that is given by the matrix of $SL(2,\mathbb{R})$

$$\begin{pmatrix} \cosh\left(\frac{k}{2}\right) & \sinh\left(\frac{k}{2}\right) \\ \sinh\left(\frac{k}{2}\right) & \cosh\left(\frac{k}{2}\right) \end{pmatrix}$$

displaces the points on this "line" by a distance of k, to the right (real positive part) if $k > 0$, and to the left if $k < 0$.

Distance between two points equidistant from a geodesic.

Figure D.1.

LEMMA D.3. *In the situation of Figure D.1, we have the formula*

$$\cosh(d(M_1, M_2)) = \frac{1}{2}\Big[\cosh(\ell) + 1 + (\cosh(\ell) - 1)\cosh(2m)\Big].$$

D.2 RELATIONS BETWEEN THE SIDES OF A RIGHT HYPERBOLIC HEXAGON

In Figure D.2, s, k, and k' are given. We want to calculate ℓ, which is the shortest distance between the lines D_1 and D_2.

LEMMA D.4. *In the above notation, we have:*

$$\cosh \ell = \cosh(s)\sinh(k)\sinh(k') - \cosh(k)\cosh(k').$$

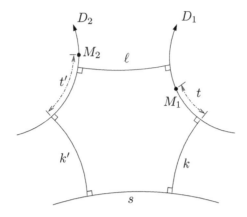

Figure D.2.

Proof. We calculate the distance from an arbitrary point M_1, of (oriented) abscissa t on D_1, to an arbitrary point M_2, of abscissa t' on D_2 (as in Figure D.2).

We place M_2 at i and we try to obtain $M_1 = f(i)$, where f is composed of hyperbolic translations along the axis iy and along the unit circle.

We easily see that one can take $f \in \mathrm{SL}(2, \mathbb{R})$ as a product:

$$f = \begin{pmatrix} e^{-t'/2} & 0 \\ 0 & e^{t'/2} \end{pmatrix} F \begin{pmatrix} e^{-t/2} & 0 \\ 0 & e^{t/2} \end{pmatrix} = \begin{pmatrix} a & b \\ c & d \end{pmatrix}$$

where F is the matrix product

$$\begin{pmatrix} \cosh\left(\frac{k'}{2}\right) & \sinh\left(\frac{k'}{2}\right) \\ \sinh\left(\frac{k'}{2}\right) & \cosh\left(\frac{k'}{2}\right) \end{pmatrix} \begin{pmatrix} e^{s/2} & 0 \\ 0 & e^{-s/2} \end{pmatrix} \begin{pmatrix} \cosh\left(\frac{k}{2}\right) & -\sinh\left(\frac{k}{2}\right) \\ -\sinh\left(\frac{k}{2}\right) & \cosh\left(\frac{k}{2}\right) \end{pmatrix}.$$

Letting the entries of F be given by $\begin{pmatrix} \alpha & \beta \\ \gamma & \delta \end{pmatrix}$, we compute

$$a = \alpha e^{-(t+t')/2} \ , \ b = \beta e^{(t-t')/2} \ , \ c = \gamma e^{(t'-t)/2} \ , \ d = \delta e^{(t+t')/2} \ ,$$

$$\alpha = e^{s/2} \cosh\left(\frac{k'}{2}\right) \cosh\left(\frac{k}{2}\right) - e^{-s/2} \sinh\left(\frac{k'}{2}\right) \sinh\left(\frac{k}{2}\right),$$

$$\beta = e^{-s/2} \sinh\left(\frac{k'}{2}\right) \cosh\left(\frac{k}{2}\right) - e^{s/2} \cosh\left(\frac{k'}{2}\right) \sinh\left(\frac{k}{2}\right),$$

$$\gamma = e^{s/2} \sinh\left(\frac{k'}{2}\right) \cosh\left(\frac{k}{2}\right) - e^{-s/2} \cosh\left(\frac{k'}{2}\right) \sinh\left(\frac{k}{2}\right),$$

$$\delta = e^{-s/2} \cosh\left(\frac{k'}{2}\right) \cosh\left(\frac{k}{2}\right) - e^{s/2} \sinh\left(\frac{k'}{2}\right) \sinh\left(\frac{k}{2}\right).$$

We have

$$\begin{aligned} 2\cosh(d(M_1, M_2)) &= a^2 + b^2 + c^2 + d^2 \\ &= \alpha^2 e^{-(t+t')} + \beta^2 e^{t-t'} + \gamma^2 e^{t'-t} + \delta^2 e^{t+t'}. \end{aligned}$$

In solving for the critical point of this function of the variables t and t' (the critical point is unique because the function is convex), we get

$$e^{t+t'} = \left|\frac{\alpha}{\delta}\right|, \qquad e^{t-t'} = \left|\frac{\gamma}{\beta}\right|.$$

The critical value is therefore $\cosh\ell = |\alpha\delta| + |\beta\gamma|$.

$$\begin{aligned} \alpha\delta &= \frac{1}{2}\big[1 + \cosh(k)\cosh(k') - \cosh(s)\sinh(k)\sinh(k')\big], \\ \beta\gamma &= \frac{1}{2}\big[\cosh(k)\cosh(k') - 1 - \cosh(s)\sinh(k)\sinh(k')\big]. \end{aligned}$$

We moreover verify that $\alpha\delta - \beta\gamma = 1$. We thus find

$$\cosh\ell = \sup\big(1, \big|\cosh(k)\cosh(k') - \cosh(s)\sinh(k)\sinh(k')\big|\big).$$

Furthermore, we see geometrically that if we start from the hexagon and augment s, we obtain a new hexagon for which ℓ is definitely nonzero ($\cosh(\ell) > 1$); moreover, ℓ is an increasing function of s (see Section 3.2). Thus

$$\cosh(\ell) = \cosh(s)\sinh(k)\sinh(k') - \cosh(k)\cosh(k').$$

$\square$

Proof. We place M_1 at i and we write $M_2 = f(i)$, where f is the following product in $\mathrm{SL}(2, \mathbb{R})$:

$$\begin{aligned} f &= \begin{pmatrix} e^{-m/2} & 0 \\ 0 & e^{m/2} \end{pmatrix} \begin{pmatrix} \cosh\left(\frac{\ell}{2}\right) & \sinh\left(\frac{\ell}{2}\right) \\ \sinh\left(\frac{\ell}{2}\right) & \cosh\left(\frac{\ell}{2}\right) \end{pmatrix} \begin{pmatrix} e^{m/2} & 0 \\ 0 & e^{-m/2} \end{pmatrix} \\ &= \begin{pmatrix} \cosh\left(\frac{\ell}{2}\right) & e^{-m}\sinh\left(\frac{\ell}{2}\right) \\ e^{m}\sinh\left(\frac{\ell}{2}\right) & \cosh\left(\frac{\ell}{2}\right) \end{pmatrix} \end{aligned}$$

We then apply Corollary D.2. $\square$

D.3 BOUNDING DISTANCES IN PAIRS OF PANTS

We consider on the pair of pants P^2 a hyperbolic metric for which the components of the boundary are geodesics of respective lengths $2m_1$, $2m_2$, $2m_3$ (attention! this is not the usual notation). Let g_{ij} be the simple geodesic orthogonal to $\partial_i P^2$ and $\partial_j P^2$; if $i = j$, it cuts P^2 into two annuli. We set $\ell_3 = \text{length}(g_{12})$, $\ell_2 = \text{length}(g_{13})$, and $\ell_1 = \text{length}(g_{23})$.

We have

$$\cosh(m_3) = \cosh(\ell_3)\sinh(m_1)\sinh(m_2) - \cosh(m_1)\cosh(m_2).$$

Thus

$$\cosh(\ell_3) = \frac{\cosh(m_3) + \cosh(m_1)\cosh(m_2)}{\sinh(m_1)\sinh(m_2)}.$$

PROPOSITION D.5. *Let M_1 (resp. M_2) be a point of abscissa $m \leq \inf(m_1, m_2)$ on $\partial_1 P^2$ (resp. $\partial_2 P^2$), where the origin is the point of intersection with g_{12} and the orientation is that of Figure D.3. For any $\epsilon > 0$, there exists a constant, depending only on ϵ, which bounds $d(M_1, M_2)$ from above, provided that m, m_1, m_2, and m_3 satisfy the inequalities (i), (ii) or (i), (iii):*

    (i) $m_1, m_2, m_3 > \varepsilon$;

    (ii) $(m_1, m_2, m_3) \in (\nabla \leq)$ and $|m| \leq (m_1 + m_2 - m_3)/2$;

    (iii) $m_1 \geq m_2 + m_3$ and $|m| \leq m_2$.

Proof. By Lemma D.3, we have to bound the quantity

$$(\cosh(\ell_3) + 1) + (\cosh(\ell_3) - 1)\cosh(2m).$$

For this, it suffices to bound $Q = [\cosh(\ell_3) - 1]\cosh(2m)$ because we have $\cosh(\ell_3) + 1 \leq Q + 2$.

Case 1: Suppose *(i)* and *(ii)* are true. We have

$$
\begin{aligned}
Q &= \left[\frac{\cosh(m_3) + \cosh(m_1 - m_2)}{\sinh(m_1)\sinh(m_2)}\right]\cosh(2m) \\
&\leq \left[\frac{\cosh(m_3) + \cosh(m_1 - m_2)}{\sinh(m_1)\sinh(m_2)}\right]\cosh(m_1 + m_2 - m_3).
\end{aligned}
$$

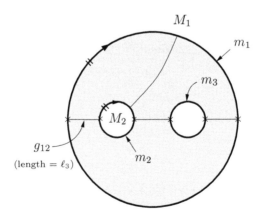

Figure D.3.

Since $\partial_1 P^2$ and $\partial_2 P^2$ play symmetric roles here, we can suppose $m_1 - m_2 \geq 0$. Then, we have

$$\cosh(m_1 - m_2)\cosh(m_1 + m_2 - m_3) \leq \cosh(2m_1 - m_3),$$
$$\cosh(m_3)\cosh(m_1 + m_2 - m_3) \leq \cosh(m_1 + m_2).$$

Further, $|m_1 - m_3| \leq m_2$. Thus $0 \leq |2m_1 - m_3| \leq m_1 + m_2$, from which we have $\cosh(2m_1 - m_3) \leq \cosh(m_1 + m_2)$.

Finally, we have

$$Q \leq \frac{2\cosh(m_1 + m_2)}{\sinh m_1 \ \sinh m_2} = 2 + 2\coth(m_1)\coth(m_2).$$

The right-hand side is bounded by *(i)*.

Case 2: Suppose *(i)* and *(iii)* are true.

$$Q = \left[\frac{\cosh(m_3) + \cosh(m_1 - m_2)}{\sinh(m_1)\sinh(m_2)}\right]\cosh(2m_2).$$

We have

$$\cosh(m_3)\cosh(2m_2) \leq \cosh(m_3 + 2m_2) \leq \cosh(m_1 + m_2),$$
$$\cosh(m_1 - m_2)\cosh(2m_2) \leq \cosh(m_1 + m_2).$$

We conclude as in the first case. $\qquad\square$

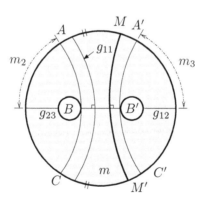

Figure D.4. All lines in this figure are geodesics

COROLLARY D.6. *Let M, M' be two distinct points of $\partial_1 P^2$ that are equidistant from the geodesic g_{11} and on the same side of it. Then $d(M, M')$ is bounded by a constant that depends only on ϵ, provided we have*

(i) $m_1, m_2, m_3 > \varepsilon$;
(ii) $m_1 \geq m_2 + m_3$;
(iii) $M \in AA'$ *(see Figure D.4).*

Proof. By the preceding proposition, the quantities $d(A, B) = d(C, B)$ and $d(A', B') = d(C', B')$ are bounded. By Lemma D.3,

$$d(M, M') \leq \sup(d(A, C), d(A', C')).$$

Then by the triangle inequality, $d(A, C) \leq 2d(A, B)$ and $d(A', C') \leq 2d(A', B')$.
□

Exposé Nine

The Classification of Surface Diffeomorphisms

by Valentin Poénaru

9.1 PRELIMINARIES

Let M be a closed, orientable surface of genus $g \geq 2$. Its compactified Teichmüller space $\overline{\mathcal{T}} = \overline{\mathcal{T}(M)}$ is homeomorphic to D^{6g-6}. The natural actions of $\pi_0(\mathrm{Diff}(M))$ on $\mathcal{T}(M)$ and on $\mathcal{PMF}(M)$ combine to give a continuous action on

$$\overline{\mathcal{T}(M)} = \mathcal{T}(M) \cup \mathcal{PMF}(M).$$

Let $\varphi \in \mathrm{Diff}(M)$ and let $[\varphi]$ be its isotopy class. By the Brouwer fixed point theorem, there is an $x \in \overline{\mathcal{T}(M)}$ such that

$$[\varphi] \cdot x = x.$$

If x belongs to $\mathcal{T}(M)$, then x determines a hyperbolic metric on M, up to isotopy, and φ is isotopic to an isometry in this metric. By Theorem 3.19, φ is isotopic to a diffeomorphism of finite order.

If x belongs to the boundary of $\overline{\mathcal{T}(M)}$, that is, $x \in \mathcal{PMF}(M)$, then the equality $[\varphi] \cdot x = x$ tells us that there exists a measured foliation whose measure class in the projective space $P(\mathbb{R}_+^{\mathcal{S}})$ is preserved by φ. In other words, there exists a measured foliation $(\mathcal{F}, \mu)$ and a scalar $\lambda \in \mathbb{R}_+$ such that

$$\varphi(\mathcal{F}, \mu) \overset{m}{\sim} (\mathcal{F}, \lambda\mu) = \lambda(\mathcal{F}, \mu). \tag{9.1}$$

Note 1. Here $\overset{m}{\sim}$ is the relation of measure equivalence between measured foliations. Recall that

$$(\mathcal{F}_1, \mu_1) \overset{m}{\sim} (\mathcal{F}_2, \mu_2)$$

means the two measured foliations define the same functional in $\mathbb{R}_+^{\mathcal{S}}$ (Schwartz equivalence). By the results of Exposé 6, this relation is the same as Whitehead equivalence, defined in Section 5.3.

Note 2. $\varphi(\mathcal{F}, \mu)$ denotes the image foliation of $\mathcal{F}$ under φ, equipped with the (direct image) measure: the measure of a transverse arc α is the μ-measure of $\varphi^{-1}(\alpha)$.

135

Now, we define a *partial measured foliation* of M as a measured foliation $(\mathcal{F}', \mu')$ that is supported on a compact submanifold N of dimension 2, and that satisfies the following:

(i) Each connected component of ∂N is a cycle of leaves.

(ii) If Γ is a component of ∂N that bounds a disk in $M \setminus \mathrm{int}\, N$, then the number of separatrices that leave the set $\mathrm{Sing}(\mathcal{F}' \cap \Gamma)$ and enter N is at least 2.

If we start with a measured foliation $(\mathcal{F}, \mu)$ of M, we may "unglue" $\mathcal{F}$ along all of the leaves that join the singularities and "blow-up" the singularities that are not connected to other singularities. We obtain, then, a partial measured foliation $U(\mathcal{F}, \mu)$, called the *unglue* of $(\mathcal{F}, \mu)$, whose singularities are all on the boundary. One easily verifies the following facts:

(a) $i_*(\mathcal{F}, \mu)$ and $i_*(U(\mathcal{F}, \mu))$ are equal in $\mathbb{R}_+^{\mathcal{S}}$.

(b) If $i_*(\mathcal{F}_1, \mu_1) = i_*(\mathcal{F}_2, \mu_2)$, that is to say, if $(\mathcal{F}_1, \mu_1) \overset{m}{\sim} (\mathcal{F}_2, \mu_2)$, then $U(\mathcal{F}_1, \mu_1)$ and $U(\mathcal{F}_2, \mu_2)$ are isotopic.

(c) Let $\beta U(\mathcal{F}, \mu)$ denote the union of the boundary components of the support of $U(\mathcal{F}, \mu)$ that do not bound a disk in M. As an element of $\mathcal{S}'$, $\beta U(\mathcal{F}, \mu)$ depends only on the measure class of $(\mathcal{F}, \mu)$.

Returning to (9.1), we have three possibilities:

(i) $\beta U(\mathcal{F}, \mu) \neq \emptyset$.

(ii) $\beta U(\mathcal{F}, \mu) = \emptyset$ and $\lambda = 1$.

(iii) $\beta U(\mathcal{F}, \mu) = \emptyset$ and $\lambda \neq 1$.

In the rest of this exposé, we will analyze the three cases. We show that *(i)* is the "reducible" case and that *(ii)* is again a case of "finite order," whereas case *(iii)* is "*pseudo-Anosov*" (see Exposé 1). The classification theorem is stated at the end of Section 9.5. In this exposé, the surfaces are always orientable, but the diffeomorphisms do not necessarily preserve orientation, which complicates certain arguments, in particular Lemma 9.9.

9.2 RATIONAL FOLIATIONS (THE REDUCIBLE CASE)

The relation (9.1) implies that $U(\varphi(\mathcal{F}, \mu))$ and $U(\mathcal{F}, \lambda\mu)$ are isotopic. Hence, in $\mathcal{S}'$, we have the equality

$$\beta U(\varphi(\mathcal{F}, \mu)) = \beta U(\mathcal{F}, \lambda\mu).$$

Further, the left-hand side is equal to $\varphi(\beta U(\mathcal{F}, \mu))$, and the right-hand side is equal to $\beta U(\mathcal{F}, \mu)$. Hence, the element $\beta U(\mathcal{F}, \mu)$ of $\mathcal{S}'$ is invariant under $[\varphi]$, with the various components possibly permuted.

Under these conditions, φ is isotopic to a diffeomorphism φ' that leaves invariant the submanifold $\beta U(\mathcal{F}, \mu)$. By cutting M along this family of curves, we obtain a manifold with boundary W, possibly disconnected, on which φ

induces a diffeomorphism ψ. We start over with an analogous study of ψ by applying Thurston's theory for surfaces with boundary, which is sketched in Exposé 11. Observe that W is simpler than M in the sense that every component of W has either smaller genus than M, or the same genus but smaller Euler characteristic in absolute value. Hence, in a finite number of stages we may determine the structure of φ up to isotopy.

9.3 ARATIONAL MEASURED FOLIATIONS

By definition, a measured foliation $(\mathcal{F}, \mu)$ is *arational* if $\beta U(\mathcal{F}, \mu)$ is empty.

LEMMA 9.1. *Let $(\mathcal{F}, \mu)$ be an arational measured foliation, and let X be the compact invariant set consisting of all singularities and all leaves joining two singularities. Then*
(1) Each connected component of X is contractible.
(2) $\mathcal{F}$ does not have any smooth closed leaves.

Proof. The manifold $\overline{M - \operatorname{Supp} U(\mathcal{F}, \mu)}$ collapses onto X, which gives *(1)*. Suppose that Γ is a smooth leaf of $(\mathcal{F}, \mu)$. Applying the stability lemma of Exposé 5 to one of the sides of Γ, one may find a maximal cylinder $\Phi \colon \Gamma \times [0,1] \to M$ such that

(i) $\Phi(\Gamma \times \{0\}) = \Gamma$.
(ii) $\Phi(\Gamma \times [0,1))$ is an embedding starting from the chosen side of Γ.

Since the genus of M is at least 2, the maximality of the cylinder implies that $\Phi(\Gamma \times \{1\}) \subset X$. In view of *(i)*, the invariant set $\Phi(\Gamma \times \{1\})$ is contractible and we may show without difficulty that $\Phi(\Gamma \times [0,1])$ is a disk D^2 with spine $\Phi(\Gamma \times \{1\})$. As there does not exist a measured foliation on D^2 where ∂D^2 is a leaf, the existence of Γ is absurd. Hence, every half-leaf of $\mathcal{F}$ that does not go to a singularity is infinite.   □

Remark. On the torus T^2, by definition, every foliation is arational, whereas a foliation that satisfies the conditions of Lemma 9.1 is conjugate to a linear foliation with irrational slope.

COROLLARY 9.2. *If $(\mathcal{F}, \mu)$ is an arational foliation, then there exists an equivalent measured foliation $(\mathcal{F}', \mu')$ that does not have any connections between singularities. This foliation is unique up to isotopy in its measure class.*

Proof. We obtain $(\mathcal{F}', \mu')$ by collapsing every component of the $\mathcal{F}$-invariant set X described above. The result of collapsing remains unchanged up to isotopy if we perform a Whitehead operation on X before collapsing; uniqueness follows. □

Convention. In what follows, we will consistently represent a class of arational foliations by the canonical model described above.

LEMMA 9.3. *If $(\mathcal{F}, \mu)$ is the canonical model of a class of arational measured foliations and if φ is a diffeomorphism such that $\varphi(\mathcal{F}, \mu) \overset{m}{\sim} \lambda(\mathcal{F}, \mu)$ for some $\lambda \in \mathbb{R}_+^*$, then φ is isotopic to φ' such that*

$$\varphi'(\mathcal{F}, \mu) = (\mathcal{F}, \lambda\mu);$$

that is to say φ' takes leaves to leaves and, for every arc α transverse to $\mathcal{F}$ we have

$$\mu(\varphi'^{-1}(\alpha)) = \lambda\mu(\alpha).$$

N.B. If $\lambda > 1$, this says that φ *contracts* the transverse distance (by a factor of $1/\lambda$), whereas if $\lambda < 1$, this says that φ *dilates* the transverse distance (by a factor of $1/\lambda$).

Proof. The foliations $\varphi(\mathcal{F}, \mu)$ and $(\mathcal{F}, \lambda\mu)$ are two canonical models of the same type; hence they are isotopic. Changing φ by this isotopy, one obtains the required φ'.   □

Let $(\mathcal{F}, \mu)$ be any measured foliation. An $(\mathcal{F}, \mu)$-*rectangle* (or, briefly, an $\mathcal{F}$-rectangle), is the image of an immersion $\varphi \colon [0,1] \times [0,1] \to M$ with the following properties:

(a) $\varphi|(0,1) \times (0,1)$ is a C^∞ embedding.

(b) $\varphi(\{t\} \times [0,1])$ is contained in a finite union of leaves and singularities; if $t \in (0,1)$, then the image is contained in a single leaf.

(c) $\varphi([0,1] \times \{0\})$ and $\varphi([0,1] \times \{1\})$ are transverse to the leaves.

For an $\mathcal{F}$-rectangle R, we consider the decomposition $\partial R = \partial_{\mathcal{F}} R \cup \partial_\tau R$ where we define

$$\partial_{\mathcal{F}} R = \varphi(\{0,1\} \times [0,1]) \quad \text{and} \quad \partial_\tau R = \varphi([0,1] \times \{0,1\}).$$

We will denote by $\partial_{\mathcal{F}}^0 R$ and $\partial_{\mathcal{F}}^1 R$ the images, respectively, of $\{0\} \times [0,1]$ and $\{1\} \times [0,1]$; an analogous notation will be used for $\partial_\tau R$. Further, we will find it convenient to write int $R = \varphi((0,1) \times (0,1))$, which in general is not the interior of the image. It is easy to see that $\text{int}(R)$ and ∂R are disjoint.

A *good system of transversals* for $\mathcal{F}$ is a finite system $\tau = \{\tau_i \mid i \in I\}$ of simple arcs with the following properties:

(a) Each arc is transverse to $\mathcal{F}$ and may meet a singularity only at one of its endpoints.

(b) Two arcs do not meet, except possibly at a single endpoint; if this is a singularity, the two arcs lie in two distinct sectors.

Remark. We do not require that every arc contains a singularity.

LEMMA 9.4. *Given a measured foliation $\mathcal{F}$ and a good system of transversals τ, there exists a unique system of rectangles $R_1, \ldots, R_N$, with the following properties:*

(1) int $R_i \cap$ int $R_j = \emptyset$ *for* $i \neq j$.

(2) $\partial_\tau^\epsilon R_i$ *is contained in a single arc of* τ, *where* $\epsilon \in \{0, 1\}$.

(3) *Each* $\partial_{\mathcal{F}}^\epsilon R_i$ *contains a point of* $\operatorname{Sing}(\mathcal{F}) \cup \partial \tau$; *in other words, every rectangle R_i is maximal with respect to condition (2).*

(4) *The two sides of each arc of* τ *are covered by the rectangles.*

Remark. It is very instructive to take a small transversal to an irrational foliation of T^2 and to construct the corresponding rectangles.

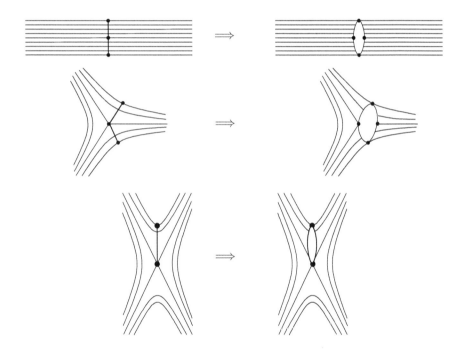

Figure 9.1.

Proof. Cut the surface along the arcs of τ as indicated in Figure 9.1. We obtain a manifold with boundary $\widehat{M}$ with a foliation $\mathcal{F}'$; the boundary τ' of $\widehat{M}$ is the "double" of τ. Consider the finite set Z of τ', defined by any one of the following conditions:

(1) $x \in \operatorname{Sing}(\mathcal{F}')$.

(2) x is one of the points giving an endpoint of τ.

(3) The leaves departing x run into a singularity of $\mathcal{F}$ or an endpoint of an arc of τ.

By Poincaré recurrence (Theorem 5.2), all leaves that depart from a point of $\tau' - Z$ return to $\tau' - Z$.

For every component α_i of $\tau' - Z$, the stability lemma (Lemma 5.4) implies that we may find a rectangle R_i such that $\partial_\tau^0 R_i = \overline{\alpha_i}$. The segment $\partial_\tau^1 R_i$ gets attached to another component of $\tau' - Z$. When we view these in M, the rectangles are the desired rectangles. Uniqueness is left as an exercise. $\square$

LEMMA 9.5. *If, in the hypotheses of Lemma 9.4, $\mathcal{F}$ is an arational foliation, then*
$$R_1 \cup \cdots \cup R_N = M.$$

Proof. The union of the R_i is a closed $\mathcal{F}$-invariant set. If the boundary is not empty, there is a closed $\mathcal{F}$-invariant set consisting of cycles of leaves. If $\mathcal{F}$ is arational, such a cycle cannot exist, hence the boundary is empty and $M = \bigcup R_i$. $\square$

LEMMA 9.6. *If $\mathcal{F}$ is an arational foliation, every half-leaf L of $\mathcal{F}$ that does not lead to a singularity is dense.*

Proof. We know that L is "infinite" (Lemma 9.1). Let τ be a small arc transverse to $\mathcal{F}$ and $R_1, \ldots, R_N$ be the system of rectangles from Lemma 9.4. By the above lemma, $\bigcup R_i$ is M and, since L is infinite, it contains plaques in $\bigcup \operatorname{int} R_i$, so L meets τ. Since τ was arbitrary, L is dense. $\square$

9.4   ARATIONAL FOLIATIONS WITH $\lambda = 1$ (THE FINITE ORDER CASE)

LEMMA 9.7. *If φ is a diffeomorphism and $(\mathcal{F}, \mu)$ is an arational foliation such that*
$$\varphi(\mathcal{F}, \mu) = (\mathcal{F}, \mu),$$
then φ is isotopic to a diffeomorphism of finite order that preserves $(\mathcal{F}, \mu)$.

Proof. In a neighborhood of each singularity, we choose transverse arcs, one in each sector, all of the same length with respect to the measure μ, as indicated in Figure 9.2.

Since $\lambda = 1$, we may choose the system of arcs τ so that, possibly after an isotopy of φ through diffeomorphisms that preserve $\mathcal{F}$, we have $\varphi(\tau) = \tau$.

Let $R_1, \ldots, R_N$ be the system of rectangles associated to τ (see Lemma 9.4). Since $\varphi(\tau) = \tau$ and $\varphi(\mathcal{F}) = \mathcal{F}$, we have that each $\varphi(R_i)$ is again an $\mathcal{F}$-rectangle satisfying condition *(2)* of Lemma 9.4. It is easy to see that there exists a permutation π of $(1, \ldots, N)$ such that $\varphi(R_i) = R_{\pi(i)}$. In particular, φ acts on the graph $\Gamma = \bigcup_i \partial R_i$ as well. Hence φ permutes the edges of Γ among themselves. Working with the cycles of this permutation we may isotope φ to

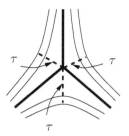

Figure 9.2.

φ', through diffeomorphisms that preserve $\mathcal{F}$, such that $\varphi'|_\Gamma$ is periodic and $\varphi'(R_i) = R_{\pi(i)}$.

Working on the cycles of π, we may make a second isotopy to obtain a periodic diffeomorphism, through diffeomorphisms that preserve $\mathcal{F}$. □

Remark. Such a diffeomorphism always has a fixed point in $\mathcal{T}(M)$. Indeed, if φ is of finite order, φ' is an isometry in a certain metric m (whose curvature we cannot control). Hence φ is an automorphism of the underlying conformal structure. By the uniformization theorem cited in Exposé 7, there is a unique hyperbolic structure underlying this structure which, as a consequence, is invariant under φ.

9.5 ARATIONAL FOLIATIONS WITH $\lambda \neq 1$ (THE PSEUDO-ANOSOV CASE)

We now suppose that we are in the situation where $\varphi(\mathcal{F}, \mu) = (\mathcal{F}, \lambda\mu)$, with $\lambda \neq 1$, where $\mathcal{F}$ is a canonical model for a class of arational foliations. By replacing φ with φ^{-1} if necessary, we may assume that $\lambda > 1$.

LEMMA 9.8. *The multiplicative factor λ (resp. $1/\lambda$) is an algebraic integer of degree bounded by a quantity that is a function only of the genus of the surface.*

Proof. There is a twofold branched covering $\widetilde{M}$ over M in which $(\mathcal{F}, \mu)$ lifts to a closed 1-form ω (the "orientation cover"). If γ is a loop of $M - \text{Sing}(\mathcal{F})$, along which $\mathcal{F}$ is orientable, then $\varphi(\gamma)$ has the same property; it follows that φ lifts to a diffeomorphism ψ of the open covering $\widetilde{M} \to M - \text{Sing}(\mathcal{F})$. This extends to a diffeomorphism $\widetilde{\psi}$ of $\widetilde{M}$. We have $(\widetilde{\varphi}^{-1})^*(\omega) = \lambda\omega$.

Hence λ is an eigenvalue of an automorphism of $H_1(\widetilde{M}, \mathbb{Z})$. Now, the rank of this cohomology group is bounded by a quantity that depends only on the genus of M. □

LEMMA 9.9. *Let* $(\mathcal{F}, \mu)$ *be as above. Up to changing* φ *by an isotopy leaving* $\mathcal{F}$ *invariant, we may find a good system of transversals* τ *with the following properties:*

(1) There is at least one arc of τ *in each sector of each singularity (Figure 9.2).*

(2) $\varphi(\tau) \subset \tau$, *that is,* φ *takes every arc of* τ *into an arc of* τ.

(3) If $x \in \partial\tau - \mathrm{Sing}(\mathcal{F})$, x *belongs to a separatrix of a singularity; we denote by* F_x *the arc of the leaf joining* x *to* $\mathrm{Sing}(\mathcal{F})$.

(4) Every separatrix contains an F_x.

(5) $\bigcup F_x \subset \varphi \left(\bigcup F_x \right)$.

Proof. Since $\lambda > 1$, φ contracts the transversals (see the definition of the direct image of a measure). Up to modifying φ by an isotopy that preserves $\mathcal{F}$, it is easy to find a good system of transversals τ'' that satisfies *(1)* and *(2)* and that has one arc in each sector. Let α'' be an arc of τ'' and L a separatrix emanating from a singularity s. Since half-leaves are dense, there is a first point of intersection of L with α'', starting from s. By considering all separatrices, we obtain on α'' a finite number of such vertices; we subdivide α'' and we truncate it at the furthest of these vertices. Let τ' be the good system of transversals obtained by this operation on each of the arcs of τ''. The system τ' satisfies *(1)*, *(3)*, *(4)*.

The system τ' also satisfies *(2)*. Let $\alpha' \in \tau'$, with endpoints x and y. The transversal α' is contained in a transversal α'' of τ'', and $\varphi(\alpha'') \subset \beta''$ for some $\beta'' \in \tau''$. We suppose for the moment that $\varphi(\alpha')$ is already contained in $\cup \{\beta' | \beta' \in \tau'\}$. If $\varphi(\alpha')$ is not contained in a single arc of τ', there exists a separatrix L where the first point of intersection with β'' is a point z between $\varphi(x)$ and $\varphi(y)$. But $\varphi^{-1}(L)$ intersects α'' in $t \neq \varphi^{-1}(z)$ before intersecting α'' in $\varphi^{-1}(z)$. Thus L intersects β'' in $\varphi(t)$, which is before z on L; this is a contradiction. Now an analogous argument proves that $\varphi(\alpha') \subset \cup \{\beta' | \beta' \in \tau'\}$ for all $\alpha' \in \tau'$, which completes the proof that τ' satisfies *(2)*.

Let n be the first nonnegative integer for which φ^{n+1} leaves invariant each separatrix. Let τ be the subdivision of τ' defined by $\tau' \vee \varphi(\tau') \vee \cdots \vee \varphi^n(\tau')$; that is, an arc α of τ is contained in an arc of τ' and is bounded by two consecutive points of the form $\varphi^j(x)$, $\varphi^{j'}(x')$, with $x, x' \in \partial\tau'$ and $0 \leq j, j' \leq n$. Properties *(1)*, *(3)*, and *(4)* are evident.

For *(2)*, we suppose that $\varphi(\alpha)$, which, since τ' satisfies *(2)*, is contained in a certain β' of τ', is subdivided. That is to say that between $\varphi^{j+1}(x)$ and $\varphi^{j'+1}(x')$, there will be a $\varphi^{j''}(x'')$, with $x'' \in \partial\tau'$, $j'' \leq n$. We claim that $j'' \geq 1$. This is true because $\varphi(\alpha)$, which is contained in β', is not subdivided by a point of $\partial\tau'$. Thus α contains $\varphi^{j''-1}(x'')$, which is a contradiction.

We now prove *(5)*. Let $x \in \partial\tau$. If $\varphi^{-1}(x) \in \partial\tau$, property *(5)* is evident. If $\varphi^{-1}(x) \notin \partial\tau$, then $x \in \partial\tau'$. The leaf L of F_x also contains $\varphi^{n+1}(F_x)$; by the same construction as τ', x is the first point of intersection of L with the arc of

τ'' that passes through x. Thus $\varphi^{n+1}(F_x)$ contains F_x and $\varphi^{-1}(F_x) \subset F_{\varphi^n(x)}$.
□

Let $x \in \partial\tau - \mathrm{Sing}\mathcal{F}$ and let L be the leaf containing F_x. Starting from the singularity s of L, we consider the first point y that belongs to τ and not to F_x. We denote by F'_x the segment from s to y on L.

Let F (resp. F') be the union of the F_x (resp. F'_x). Seeing that $\varphi(\tau) \subset \tau$ and $\varphi(F) \supset F$, we verify without difficulty that $\varphi(F') \supset F'$.

Let $(\mathcal{F}, \mu)$ be an arational measured foliation and φ a diffeomorphism such that $\varphi(\mathcal{F}, \mu) = \lambda(\mathcal{F}, \mu)$ with $\lambda > 1$. A *pre-Markov partition* for $(\mathcal{F}, \varphi)$ is by definition a collection of $\mathcal{F}$-rectangles $R_1, \ldots, R_m$ such that:

(1) int $R_i \cap$ int $R_j = \emptyset$.

(2) $\cup R_i = M$.

(3) $\varphi(\cup\partial_\tau R_i) \subset \cup\partial_\tau R_i$.

(4) $\varphi^{-1}(\cup\partial_\mathcal{F} R_i) \subset \cup\partial_\mathcal{F} R_i$.

LEMMA 9.10. *A pre-Markov partition also satisfies*

(5) For each $i = 1, \ldots, m$ and $\epsilon \in \{0, 1\}$, $\varphi(\partial_\tau^\epsilon R_i)$ is covered on the side corresponding to $\varphi(R_i)$ by a single rectangle: $\varphi(\partial_\tau^\epsilon R_i) \subset \partial_\tau^{\eta(\epsilon,i)} R_{j(\epsilon,i)}$.

(6) Similarly, $\varphi^{-1}(\partial_\mathcal{F}^\epsilon R_i)$ is covered on the side corresponding to $\varphi^{-1}(R_i)$ by a single rectangle.

This is to say that the image under φ of a rectangle R_i is something like Figure 9.3.

Proof. If *(5)* does not hold, there exists an $x \in$ int R_i such that $\varphi(x) \in \partial_\mathcal{F} R_j$, which contradicts *(4)*. Similarly, *(6)* follows from *(3)*.   □

LEMMA 9.11. *Let $(\mathcal{F}, \mu)$ be the canonical model of a class of arational foliations and let φ be a diffeomorphism such that $\varphi(\mathcal{F}, \mu) = (\mathcal{F}, \lambda\mu)$ with $\lambda > 1$. After possibly performing an isotopy of φ preserving $\mathcal{F}$, there exists a pre-Markov partition for $(\mathcal{F}, \varphi)$.*

Proof. Let τ be as in Lemma 9.9 and let $R'_1, \ldots, R'_l$ be the system of rectangles obtained from τ. We construct the system $R_1, \ldots, R_m$ by taking the closures of the components of $\cup$ int $R'_i - F'$, where F' is described in the remark following Lemma 9.9.

Conditions *(1)* and *(2)* for a pre-Markov partition are clearly satisfied. For condition *(3)*, we see that $\tau = \cup\partial_\tau R'_i = \cup\partial_\tau R_i$ and we know that $\varphi(\tau) \subset \tau$. Moreover, by construction, each $\partial_\mathcal{F}^\epsilon R'_i$ is an arc α of a leaf that joins a point $x \in \partial\tau$ to a point $y \in \tau$ and does not intersect τ in its interior. If $x \notin \mathrm{Sing}\,\mathcal{F}$, then α is contained in F'_x. If x is a singularity, then α is contained in F_y. Thus $\partial_\mathcal{F}^\epsilon R'_i \subset F'$ in all cases. The subdivision guarantees that F' is covered by the union of the $\partial_\mathcal{F} R_i$. We have remarked that $\varphi^{-1}(F') \subset F'$, and so condition *(4)* is satisfied.   □

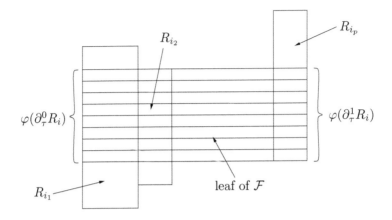

(N.B.: $i_1, i_2, \ldots, i_p$ are not necessarily distinct)

Figure 9.3. The image under φ of a rectangle R_i

In the rest of this exposé, we work with a pre-Markov partition $R_1, \ldots, R_m$ adapted to the measured foliation $(\mathcal{F}, \mu)$ and to the diffeomorphism φ. We denote by x_i the μ-length of the rectangle R_i and by a_{ij} the number of times that $\varphi(\mathrm{int}\, R_i)$ crosses $\mathrm{int}\, R_j$ (i.e., the number of components of the intersection). Since φ^{-1} dilates transverse distances by a factor of λ and since a_{ij} is also equal to the number of times that $\varphi^{-1}(\mathrm{int}\, R_j)$ crosses $\mathrm{int}\, R_i$, we find

$$\lambda x_j = \sum_i x_i a_{ij}.$$

In other words, the column vector x_i is an eigenvector, with eigenvalue λ, for the transpose matrix of $A = (a_{ij})$.

LEMMA 9.12. *There exist numbers $\xi > 0$ and $y_1, \ldots, y_m > 0$ such that*

$$y_i = \xi \sum_j a_{ij} y_j.$$

In other words, A admits an eigenvalue $\xi^{-1} > 0$, with an eigenvector whose coordinates are all strictly positive.

Proof. Since $a_{ij} \geq 0$ and, for each j, there exists an i such that $a_{ij} > 0$, A acts projectively on the fundamental simplex. The Brouwer fixed point theorem then implies that A has a positive eigenvalue with an eigenvector $(y_1, \ldots, y_n)$, where each y_i is nonnegative and $\sum y_i > 0$. It suffices to show that, for all i, y_i is nonzero.

Let us say, to fix notation, that $y_1 = y_2 = \cdots = y_l = 0$ and that $y_{l+1} > 0, \ldots, y_n > 0$. It follows that, for $i \leq l$, we have

$$a_{ij} > 0 \Rightarrow j \leq l.$$

In other words, the set

$$J = \bigcup_{i=1}^{l} R_i$$

is invariant under φ and is not dense. To show that this is a contradiction, we can make the following remarks.

1. First of all, for any integer $N > 0$, we have that $R_1, \ldots, R_m$ is a pre-Markov partition for φ^N. Thus, without loss of generality, we can reduce to the case where φ fixes each sector of each singular point (in particular φ^N fixes each singularity).

2. As $\varphi(J) \subset J$ and as each segment of τ is contracted by φ toward its singular point, there exists among the rectangles $R_1, \ldots, R_l$, a rectangle, let us say R_1, which is in the configuration shown in Figure 9.4.

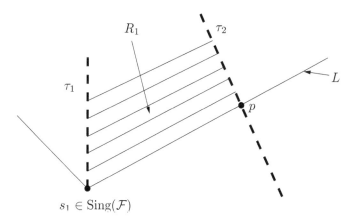

Figure 9.4.

3. Since τ_2 is contracted by φ toward its singularity, the points $\varphi^n(p)$ form an infinite set. Further, they all belong to the same leaf L which is φ-invariant. Thus the sequence converges toward infinity in the topology of the leaf. If F denotes the segment from s_1 to p along L, we have

$$L = \bigcup_{n=1}^{\infty} \varphi^n(F).$$

4. As the leaf L is dense in M, the preceding equality implies that

$$\bigcup_{n=1}^{\infty} \varphi^n(R_1)$$

is dense in M. On the other hand, this union is contained in J which is not dense; this is a contradiction.   □

Construction of a measured foliation $\mathcal{F}'$. We are going to construct a measured foliation $\mathcal{F}'$ that has the same singularities as $\mathcal{F}$, is transverse to $\mathcal{F}$ outside of these points, and satisfies the following properties:

(A) Each segment of τ is contained in a leaf of $\mathcal{F}'$ and each rectangle R_i is foliated by $\mathcal{F}'$, as in Figure 9.5.

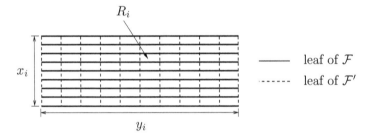

R_i

x_i

y_i

—— leaf of $\mathcal{F}$

------ leaf of $\mathcal{F}'$

Figure 9.5.

(B) The $\mathcal{F}'$-width of R_i is the y_i of Lemma 9.12.

(C) Let $A_0, A_1, \ldots, A_k$ be the sequence of vertices of τ found on $\partial_{\mathcal{F}}^0 R_i$. The segments $[A_1, A_2], \ldots, [A_{k-2}, A_{k-1}]$ are all of the type $\partial_{\mathcal{F}}^\epsilon R_j$. Thus their $\mathcal{F}'$-widths are prescribed by condition *(B)*. Let u be the $\mathcal{F}'$-width of $[A_0, A_1]$. We determine u in the following way: if q is a large enough integer, then $\varphi^q(A_0)$ and $\varphi^q(A_1)$ do not belong to $\partial \tau - \mathrm{Sing}(\mathcal{F})$. Thus $\varphi^q([A_0, A_1])$ is a sum of segments (plaques), contained in a finite union of leaves and singularities, each traversing an R_j from side to side. We claim then

$$u = \xi^q \times (\text{sum of the } \mathcal{F}'\text{-widths of these segments}).$$

We do the same for $[A_{k-1}, A_k]$, and for the intervals of $\partial_{\mathcal{F}}^1 R_i$. The lemma below says that all these choices are consistent.

LEMMA 9.13. *The* $\mathcal{F}'$-*width of* $[A_0, A_k]$ *is equal to the sum of the* $\mathcal{F}'$-*widths of the* $[A_j, A_{j+1}]$, $j = 0, \ldots, k - 1$.

Proof. To fix notation, let us say that $\partial_{\mathcal{F}}^0 R_i$ is $[A_0, A_1] \cup [A_1, A_2] \cup [A_2, A_3]$ where $[A_1, A_2]$ is a plaque of R_1 and where $[A_2, A_3]$ is a plaque of R_2; also, say that $\varphi([A_0, A_1])$ is the sum of a plaque of R_3 and a plaque of R_4. It suffices to prove that

$$y_i = \xi y_3 + \xi y_4 + y_1 + y_2.$$

We set

$$a_{ij}^{(q)} = \operatorname{card} \pi_0(\varphi^q(\operatorname{int} R_i) \cap \operatorname{int} R_j)$$

and

$$a_{ij}^q = \operatorname{card} \pi_0(\varphi^q(\partial_{\mathcal{F}}^0 R_i) \cap R_j).$$

We remark that $a_{ij}^{(q)}$ is the (i, j)-entry of the matrix A^q; further, the integer $\delta_{ij}^q = a_{ij}^q - a_{ij}^{(q)}$ is between 0 and 4 inclusive. Indeed, $\varphi^q(\partial_{\mathcal{F}}^0 R_i) \cap R_j$ can contain two arcs of $\partial_{\mathcal{F}} R_j$, as well as the two images of the endpoints of $\partial_{\mathcal{F}}^0 R_j$, in addition to the intersections of the interiors.

Finally, if q is big enough so that $\varphi^q(\partial \tau) \cap \partial \tau = \emptyset$, we have the equality given by the geometry

$$a_{ij}^q = a_{3j}^{q-1} + a_{4j}^{q-1} + a_{ij}^q + a_{2j}^q.$$

Thus

$$\sum_j \left[a_{ij}^{(q)} - a_{3j}^{(q-1)} - a_{4j}^{(q-1)} - a_{1j}^{(q)} - a_{2j}^{(q)} \right] y_j$$

$$= \sum_j \left[\delta_{3j}^{q-1} + \delta_{4j}^{q-1} + \delta_{1j}^q + \delta_{2j}^q - \delta_{ij}^q \right] y_j.$$

The left-hand side is equal to

$$[y_i - \xi y_3 - \xi y_4 - y_1 - y_2]/\xi^q.$$

On the other hand, the right-hand side takes only a finite number of values as q varies. This forces the numerator above to be zero.   □

We provide each rectangle R_i with a system of coordinates X^i, Y^i such that

$$R_i = \{0 \leq X^i \leq x_i, \ 0 \leq Y^i \leq y_i\},$$

and such that, for each segment $[A_j, A_{j+1}]$ of $\partial_{\mathcal{F}} R_i$, the difference $Y^i(A_{j+1}) - Y^i(A_j)$ is the width prescribed by condition *(C)*. We can thus interpret the rectangles as being an atlas that defines the foliation $(\mathcal{F}', \mu')$, where the plaques are

$$Y = \text{constant}$$

and where the transverse measure is $\mu' = |dY|$.

Construction of a "diffeomorphism."   We are going to construct a "diffeomorphism" φ', isotopic to φ, such that $\varphi'(\mathcal{F}, \mu) = (\mathcal{F}, \lambda\mu)$ and $\varphi'(\mathcal{F}', \mu') = (\mathcal{F}', \xi\mu')$.

Actually, φ' is going to be a diffeomorphism on the complement of the singularities, but will not be C^1 at the singularities (see the definition of a pseudo-Anosov diffeomorphism below).

We define φ' by the following conditions:

(α) $\varphi'(R_i) = \varphi(R_i)$.

(β) $\varphi'(X^i = \text{constant}) = \varphi(X^i = \text{constant})$.

(γ) Let V be a component of $R_i \cap \varphi^{-1}(R_j)$; then

$$\varphi'(V \cap (Y^i = \text{constant})) \subset (Y^j = \text{constant}).$$

(δ) For $p, q \in V$, we have

$$\xi \, |Y^j(\varphi'(p)) - Y^j(\varphi'(q))| = |Y^i(p) - Y^i(q)|.$$

LEMMA 9.14. *We have* $\xi = 1/\lambda$.

Proof. Once $(\mathcal{F}, \mu)$ and $(\mathcal{F}', \mu')$ are given, and are transverse to each other, we have a measure $\mathcal{M}$ on M given locally by the product of μ and μ'. Clearly

$$\varphi_\star \mathcal{M} = \lambda \xi \mathcal{M}.$$

As M is compact, $\mathcal{M}$ is of finite total measure and the above equality is only possible if $\lambda \xi = 1$.   □

Remark 1. The measure $\mathcal{M}$ is thus φ'-invariant. We can show that $(\mathcal{M}, \varphi')$ is a Bernoulli process. In particular, $(\mathcal{M}, \varphi')$ is *ergodic* (see Section 10.6).

Remark 2. We note the contrast between the fact that there does not exist a compact space X equipped with a measure $\mathcal{M}$ and a homeomorphism ψ such that $\psi_\star \mathcal{M} = \lambda \mathcal{M}$, $\lambda \neq 1$, and the fact that there exists a compact manifold, equipped with a measured foliation $(\mathcal{F}, \mu)$ and a diffeomorphism $\varphi : M \to M$ such that $\varphi(\mathcal{F}, \mu) = (\mathcal{F}, \lambda\mu)$, $\lambda \neq 1$.

The set of leaves of $\mathcal{F}$ endowed with the measure μ can be seen as a "noncommutative space," and the homeomorphism of this space induced by φ satisfies the above equation. However, such a "paradoxical" situation can happen only for a "discrete" set of values λ, satisfying certain "arithmetic" conditions.

Pseudo-Anosov diffeomorphisms. By definition, a homeomorphism $\varphi : M \to M$ is a *pseudo-Anosov diffeomorphism* if there are two mutually transverse invariant measured foliations $(\mathcal{F}^s, \mu^s)$ and $(\mathcal{F}^u, \mu^u)$ and a $\lambda > 1$, such that

$$\begin{aligned} \varphi(\mathcal{F}^s, \mu^s) &= (\mathcal{F}^s, \frac{1}{\lambda}\mu^s), \\ \varphi(\mathcal{F}^u, \mu^u) &= (\mathcal{F}^u, \lambda\mu^u). \end{aligned}$$

We call $\mathcal{F}^s$ and $\mathcal{F}^u$ the *stable* and *unstable* foliations, respectively. The homeomorphism is contracting on the leaves of the stable foliation, where the lengths are measured by μ^u.

From the point of view of smoothness, a pseudo-Anosov diffeomorphism is a true diffeomorphism on $M - \text{Sing}(\mathcal{F})$; but it is never C^1 at the singularities. Of course, at the singularities, there are, topologically speaking, canonical local models of the pseudo-Anosov, coming from quadratic differentials.

We have just proven the following lemma.

LEMMA 9.15. *If the diffeomorphism φ satisfies the conditions of Section 9.1 case (iii), that is, if there exists an arational measured foliation $(\mathcal{F}, \mu)$ and a $\lambda \neq 1$ such that $\varphi(\mathcal{F}, \mu) \overset{m}{\sim} (\mathcal{F}, \lambda\mu)$, then φ is isotopic to a pseudo-Anosov diffeomorphism.*

Remark. The hypothesis that $\mathcal{F}$ is arational is essential. Indeed, the existence of pseudo-Anosov diffeomorphisms on a manifold with boundary implies that, on a closed surface M of genus ≥ 2, there exists a measured foliation $(\mathcal{F}, \mu)$ having a cycle of leaves and a diffeomorphism φ satisfying $\varphi(\mathcal{F}, \mu) \overset{m}{\sim} (\mathcal{F}, \lambda\mu)$ with $\lambda \neq 1$. As we will see, this φ is not isotopic to a pseudo-Anosov diffeomorphism of M.

We say that a homeomorphism of a surface is *reducible* if it fixes a system of simple curves that are mutually disjoint and not homotopic to a point.

THEOREM 9.16 (Classification of surface diffeomorphisms). *Let φ be a diffeomorphism of a surface of genus at least 2. Up to isotopy, φ is of one of the following types:*

(1) an isometry for a hyperbolic structure;

(2) reducible;

(3) pseudo-Anosov.

Moreover, (3) is mutually exclusive to both (1) and (2).

Once we have extended the theory to the case of nonempty boundary (Exposé 11), we will be able to say that there is a homeomorphism φ' isotopic to φ and a decomposition of M as a union of subsurfaces with boundary $M = M_1 \cup \cdots \cup M_n$, with the interiors of the M_i disjoint, such that $\varphi'(M_i) = M_i$ and that $\varphi'|_{M_i}$ is isotopic (as a homeomorphism of M_i, the boundary being free) to a hyperbolic isometry or to a pseudo-Anosov diffeomorphism. Of course, this decomposition cannot take into account any Dehn twists that φ' does on the curves along which the M_i are glued.

Proof. The classification is completely proven; it remains to prove the exclusions.

The equality $\varphi(\mathcal{F}, \mu) = (\mathcal{F}, \lambda\mu)$ with $\lambda \neq 1$ prohibits the isotopy class of φ from being periodic; thus, we obtain the incompatibility of *(1)* and *(3)*.

We suppose that φ fixes an element of $\mathcal{S}'$; up to replacing φ by one of its powers, we can suppose that φ is pseudo-Anosov and preserves the isotopy class of a curve γ. We thus have

$$I(\mathcal{F}^s, \mu^s; [\gamma]) = \lambda I(\mathcal{F}^s, \mu^s; [\varphi(\gamma)]) = \lambda I(\mathcal{F}^s, \mu^s; [\gamma]).$$

We deduce that $I(\mathcal{F}^s, \mu^s; [\gamma]) = 0$. By Proposition 5.9, there is a foliation equivalent to $\mathcal{F}^s$ with a nontrival cycle of leaves. Since $\mathcal{F}^s$ is arational, this is a contradiction. □

Remark. The incompatibility relations are in fact consequences of the dynamics of a pseudo-Anosov diffeomorphism on the compactification of Teichmüller space: there are only two fixed points, represented by the stable and unstable foliations, which are respectively attracting and repelling (see Exposé 12).

9.6   SOME PROPERTIES OF PSEUDO-ANOSOV DIFFEOMORPHISMS

LEMMA 9.17. *The stable and unstable foliations of a pseudo-Anosov diffeomorphism do not have connections between singularities; they are thus canonical models for the classes of arational foliations.*

Proof. Considering the case where the diffeomorphism fixes each singularity, such a connection must be contracted or dilated, which is impossible. □

PROPOSITION 9.18. *If U is a nonempty open set that is invariant under a pseudo-Anosov diffeomorphism, then U is dense.*

Proof. It suffices to consider the case where the diffeomorphism φ fixes the singularities of the stable and unstable foliations. Let F be a separatrix of the stable foliation, emanating from a singularity s. Since F is dense in M, there exists a segment J of F contained in U. Let a and b be the endpoints of J. The sequences $\varphi^n(a)$ and $\varphi^n(b)$ converge to s. Let T be a plaque of $\mathcal{F}^u$ that is contained in U and intersects J in one point. As we increase n, the arc $\varphi^n(T)$ is lengthened from the point of view of the transverse measure of the stable foliation and approaches the two separatrices F' and F'' adjacent to F (Figure 9.6). Precisely, $F' \cup F''$ is contained in the closure of

$$\bigcup_{n \geq 0} \varphi^n(T).$$

Thus $\overline{U}$ contains a separatrix of $\mathcal{F}^u$ entirely. A separatrix is dense, therefore $\overline{U}$ is dense. □

COROLLARY 9.19. *A pseudo-Anosov diffeomorphism is topologically transitive; that is, there exists a dense orbit.*

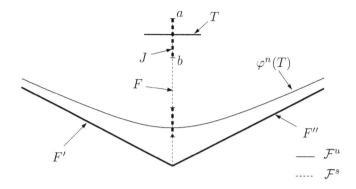

Figure 9.6. $\varphi^n(T)$ lengthens and approaches the two separatrices adjacent to F

Proof. Let $\{U_i\}$ be a countable basis of open sets for the surface. The intersection

$$\bigcap_i \left(\bigcup_{n \in \mathbb{Z}} \varphi^n(U_i) \right)$$

is nonempty by the Baire category theorem. Each point of the intersection has a dense orbit.   □

PROPOSITION 9.20. *The periodic points of a pseudo-Anosov diffeomorphism are dense.*

This is a generalization of the analogous classical fact for Anosov diffeomorphisms.

Proof. The singular points are periodic. Let $x_0 \in M$ be a regular point and let U be a rectangle that is adapted to the foliations $\mathcal{F}^s$ and $\mathcal{F}^u$ and that is a neighborhood of x_0. Let V be another rectangle neighborhood of x_0 that is strongly included in U (Figure 9.7). As the diffeomorphism φ leaves invariant a measure that assigns a nonzero measure to each nonempty open set (the measure is given locally by the product of μ^s and μ^u), Poincaré recurrence [Sin76, p. 7] applies: for any n_0, there exists an $n \geq n_0$ such that $\varphi^n(V) \cap V \neq \emptyset$.

Let x_1 be a point of V such that $\varphi^n(x_1) \in V$. Let J be the $\mathcal{F}^s$-plaque of U passing through x_1. We have

$$\mu^u(\varphi^n(J)) = \lambda^{-n}\mu^u(J)$$

where λ is the dilatation factor of φ. We see that if n_0 is chosen to be large enough (U and V being given), we will be able to ensure that $\varphi^n(J)$ is contained in U.

Identifying $\varphi^n(J)$ to an interval of J (the identification being given by following the $\mathcal{F}^u$-plaques) we see that φ^n has a "fixed point in J," which is to say that there exists a point x_2 of J where the $\mathcal{F}^u$-leaf is invariant by φ^n.

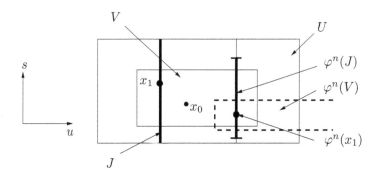

Figure 9.7. A rectangle neighborhood V of x_0 strongly included in U

Let L be the $\mathcal{F}^u$-plaque of x_2; if n_0 is chosen to be large enough, we can be sure that $\varphi^n(L)$ contains L, since $\varphi^n(L)$ and L already have $\varphi^n(x_2)$ in common (this new condition on n_0 depends only on the μ^s-widths of U and V). Thus, there is a fixed point for $\varphi^n|_L$.   □

PROPOSITION 9.21. *Let ρ be a Riemannian metric on M and $\alpha \in \mathcal{S}$. Denote by $l_\rho(\alpha)$ the length of a minimizing geodesic of the class α. Let φ be a pseudo-Anosov diffeomorphism of M of dilatation $\lambda > 1$; the isotopy class of $\varphi(\alpha)$ is well-defined. We have*

$$\lim_{n \to \infty} \sqrt[n]{l_\rho(\varphi^n(\alpha))} = \lambda.$$

Proof. If $(\mathcal{F}^s, \mu^s)$ and $(\mathcal{F}^u, \mu^u)$ are the stable and unstable foliations of φ, we can define the metric $\mu = \sqrt{(\mu^s)^2 + (\mu^u)^2}$. This metric comes from a singular norm on the tangent bundle, where the zeros are the singularities of the invariant foliations. We note in passing that the metric μ is flat in the complement of the singularities and that the curvature is constituted of Dirac masses at the singularities. Let c be a curve in the class α. We have

$$I(\mathcal{F}^s, \mu^s; \alpha) \le l_\mu(\alpha) \le \int_c d\mu^s + \int_c d\mu^u$$

and

$$I(\mathcal{F}^s, \mu^s; \varphi^n(\alpha)) \le l_\mu(\varphi^n(\alpha)) \le \int_{\varphi^n(c)} d\mu^s + \int_{\varphi^n(c)} d\mu^u.$$

By the properties of φ, it follows that

$$\lambda^n I(\mathcal{F}^s, \mu^s; \alpha) \le l_\mu(\varphi^n(\alpha)) \le \lambda^n \int_c d\mu^s + \lambda^{-n} \int_c d\mu^u.$$

As we saw in the proof of the classification theorem, $I(\mathcal{F}^s, \mu^s; \alpha) \ne 0$.

Therefore $\lim \sqrt[n]{l_\mu(\varphi^n(\alpha))} = \lambda$. The proposition follows from Lemma 9.22 below.   □

LEMMA 9.22. *Let ρ be any metric on M, and let μ be the singular metric coming from the stable and unstable foliations for a pseudo-Anosov diffeomorphism φ of M. There exist constants $K, k > 0$ such that, for any class of loops α, we have*

$$k \leq \frac{l_\rho(\alpha)}{l_\mu(\alpha)} \leq K.$$

Proof (A. Douady). Let $a_1, \ldots, a_q$ be the singularities of μ. Let $D(a, r)$ be the disk of radius r centered at a with respect to the metric μ. We choose r small enough so that the disks $D(a_i, 2r)$ are disjoint. In the complement of the disks of radius $r/2$, the two metrics give norms on the tangent bundle. Thus there exist constants $K', k' > 0$ such that, for all rectifiable *arcs* β, we have

$$k' \leq \frac{L_\rho(\beta)}{L_\mu(\beta)} \leq K', \tag{9.2}$$

where L_ρ (resp. L_μ) denotes the geometric length.

Moreover, there exist constants $K'', k'' > 0$ such that, for all $x, y \in \partial D(a_i, r)$, we have

$$k'' \leq \frac{d_\rho(x, y)}{d_\mu(x, y)} \leq K''. \tag{9.3}$$

Now, if x and y are close enough, the inequality (9.2) applies. On the other hand, if (x, y) is outside some fixed neighborhood of the diagonal, the quotient above is defined, continuous and positive on a compact set. In either case, the inequality (9.3) is clear.

We take $k = \inf(k', k'', 1)$ and $K = \sup(K', K'', 1)$. These constants depend on the choice of radius r. We take r small enough so that, for all $x, y \in \partial D(a_i, r)$, the shortest ρ-geodesic joining x to y is the identity in $\pi_1(M, D(a_i, r))$.

Let c_1 be a minimizing μ-geodesic of the class α and let c_1' be the loop obtained by replacing each diagonal of c_1 in the disks $D(a_i, r)$ by the ρ-geodesic joining the entry point to the exit point (a diagonal is a connected component of $c_1^{-1}(D)$). We thus have

$$kl_\mu(\alpha) \leq L_\rho(c_1') \leq Kl_\mu(\alpha).$$

From this we deduce that $l_\rho(\alpha) \leq Kl_\mu(\alpha)$. To obtain the other inequality, we start from a minimizing ρ-geodesic c_2 and we replace its diagonals in the disks by μ-geodesic arcs. $\square$

Exposé Ten

Some Dynamics of Pseudo-Anosov Diffeomorphisms

by Albert Fathi and Michael Shub

We prove in this exposé that a pseudo-Anosov diffeomorphism realizes the minimum topological entropy in its isotopy class. In Section 10.1 we define topological entropy and give its elementary properties. In Section 10.2 we define the growth of an endomorphism of a group and show that the topological entropy of a map is greater than the growth of the endomorphism it induces on the fundamental group. In Section 10.3, we define subshifts of finite type and give some of their properties. In Section 10.4, we prove that the topological entropy of a pseudo-Anosov diffeomorphism is the growth rate of the automorphism induced on the fundamental group; it is also $\log \lambda$, where $\lambda > 1$ is the stretching factor of f on the unstable foliation. In Section 10.5, we prove the existence of a Markov partition for a pseudo-Anosov diffeomorphism; this fact is used in Section 10.4. In Section 10.6, we show that a pseudo-Anosov map is Bernoulli.

10.1 TOPOLOGICAL ENTROPY

Topological entropy was defined to be a generalization of measure theoretic entropy [AKM65]. In some sense, entropy is a number (possibly infinite) that describes "how much" dynamics a map has. Here the emphasis, of course, must be on asymptotic behavior. For example, if $f\colon X \to X$ is a map and $N_n(f)$ is the cardinality of the fixed point set of f^n, then $\limsup \frac{1}{n} \log N_n(f)$ is one measure of "how much" dynamics f has. However, if we consider

$$f \times R_\theta \colon X \times T^1 \to X \times T^1$$

to be

$$(f \times R_\theta)(x, \alpha) = (f(x), \theta + \alpha)$$

where $T^1 = \mathbb{R}/\mathbb{Z}$ and θ is irrational, then $N_n(f \times R_\theta) = 0$, and yet $f \times R_\theta$ should have at least as "much" dynamics as f. Topological entropy is a topological invariant that overcomes this difficulty.

We describe a lot of material frequently without crediting authors.

Topological entropy. Let $f\colon X \to X$ be a continuous map of a compact topological space X. Let $\mathcal{A} = \{A_i\}_{i \in I}$ and $\mathcal{B} = \{B_j\}_{j \in J}$ be open covers of X. The open cover $\{A_i \cap B_j\}_{i \in I, j \in J}$ will be denoted by $\mathcal{A} \vee \mathcal{B}$. If $\mathcal{A}$ is a cover, $N_n(f, \mathcal{A})$ denotes the minimum cardinality of a subcover of $\mathcal{A} \vee f^{-1}(\mathcal{A}) \vee \cdots \vee f^{-n+1}(\mathcal{A})$, and $h(f, \mathcal{A}) = \limsup \frac{1}{n} \log N_n(f, \mathcal{A})$. The *topological entropy* of f is

$$h(f) = \sup_{\mathcal{A}} h(f, \mathcal{A})$$

where the supremum is taken over all open covers of X.

PROPOSITION 10.1. *Let X and Y be compact spaces. Let $f \colon X \to X$, $g \colon Y \to Y$, and $p \colon X \to Y$ be continuous. Suppose that p is surjective and $pf = gp$:*

$$
\begin{array}{ccc}
X & \xrightarrow{\ f\ } & X \\
{\scriptstyle p}\big\downarrow & & \big\downarrow{\scriptstyle p} \\
Y & \xrightarrow{\ g\ } & Y
\end{array}
$$

then $h(f) \geq h(g)$.

In particular, if p is a homeomorphism, then $h(f) = h(g)$. So topological entropy is a topological invariant.

Proof. Pull back the open covers of Y to open covers of X. ☐

For metric spaces, compact or not, Bowen has proposed the following definition of topological entropy:

Suppose $f \colon X \to X$ is a continuous map of a metric space X and suppose $K \subset X$ is compact. Let $\epsilon > 0$.

- We say that a set $E \subset K$ is (n, ϵ)-*separated* if, given $x, y \in E$ with $x \neq y$, there is $0 \leq i < n$ such that $d(f^i(x), f^i(y)) \geq \epsilon$. We let $s_K(n, \epsilon)$ be the maximal cardinality of an (n, ϵ)-separated set contained in K.

- We say that the set E is (n, ϵ)-*spanning* for K if, given $y \in K$, there is an $x \in E$ such that $d(f^i(x), f^i(y)) < \epsilon$ for each i with $0 \leq i < n$. We let $r_K(n, \epsilon)$ be the minimal cardinality of an (n, ϵ)-spanning set contained in K.

It is easy to see that

$$r_K(n, \epsilon) \leq s_K(n, \epsilon) \leq r_K(n, \epsilon/2).$$

We let

$$
\bar{s}_K(\epsilon) \;=\; \limsup \frac{1}{n} \log s_k(n, \epsilon) \quad \text{and}
$$

$$
\bar{r}_K(\epsilon) \;=\; \limsup \frac{1}{n} \log r_k(n, \epsilon).
$$

Obviously $\bar{s}_K(\epsilon)$ and $\bar{r}_K(\epsilon)$ are decreasing functions of ϵ, and

$$\bar{r}_K(\epsilon) \leq \bar{s}_K(\epsilon) \leq \bar{r}_K(\epsilon/2).$$

Hence, we may define $h_K(f) = \lim_{\epsilon \to 0} \bar{s}_K(\epsilon) = \lim_{\epsilon \to 0} \bar{r}_K(\epsilon)$. Finally, we set

$$h_X(f) = \sup\{h_K(f) \mid K \subset X \text{ compact}\}.$$

PROPOSITION 10.2. [Bow71, Din71]. *If X is a compact metric space and $f\colon X \to X$ is continuous, then $h_X(f) = h(f)$.*

The proof is rather straightforward. By the Lebesgue covering lemma, every open cover has a refinement that consists of ϵ-balls.

The number $h_X(f)$ depends on the metric on X and makes best sense for uniformly continuous maps.

Suppose that X and Y are metric spaces, we say that $p\colon X \to Y$ is a *metric covering map* if it is surjective and satisfies the following condition: there exists $\epsilon > 0$ such that, for any $0 < \delta < \epsilon$, any $y \in Y$ and any $x \in p^{-1}(y)$, the map $p\colon B_\delta(x) \to B_\delta(y)$ is a bijective isometry (here $B_\delta(\cdot)$ is the δ-ball).

The main example we have in mind is the universal covering $p\colon \widetilde{M} \to M$ of a compact differentiable manifold M.

PROPOSITION 10.3. *Suppose that $p\colon X \to Y$ is a metric covering and $f\colon X \to X$, $g\colon Y \to Y$ are uniformly continuous. If $pf = gp$, then $h_X(f) = h_Y(g)$.*

Proof. It should be an easy estimate. The clue is that for $\ell > 0$ and for any sequence a_n we have $\limsup \frac{1}{n} \log(\ell a_n) = \limsup \frac{1}{n} \log a_n$. If $K \subset X$ and $K' \subset Y$ are compact and $p(K) = K'$, then there is a number $\ell > 0$ such that $\text{card}(p^{-1}(y) \cap K) \leq \ell$ for all $y \in K'$. In fact, we may choose ℓ such that if $\delta > 0$ is small enough, then $p^{-1}(B_\delta(y)) \cap K$ can be covered by at most ℓ 2δ-balls centered at points in $p^{-1}(B_\delta(y)) \cap K$.

By the uniform continuity of f, we can find a δ_0 ($< \epsilon$) such that $x, x' \in X$ and $d(x, x') < \delta_0$ implies $d(f(x), f(x')) < \epsilon$, where $\epsilon > 0$ is the one given in the definition of a metric covering. If $2\delta < \delta_0$, it is easy to see that if $E' \subset K'$ is an (n, δ)-spanning set for g, then there exists an $(n, 2\delta)$-spanning set $E \subset K$ for f, such that $\text{card } E \leq \ell \, \text{card } E'$. So, we have $r_K(n, 2\delta) \leq \ell r_{K'}(n, \delta)$, hence $\bar{r}_K(f, 2\delta) \leq \bar{r}_{K'}(g, \delta)$ and $h_K(f) \leq h_{K'}(g)$.

On the other hand, if $E \subset K$ is (n, η)-spanning (with $0 < \eta < \epsilon$) then $p(E) \subset K'$ is (n, η)-spanning. So $r_{K'}(n, \eta) \leq r_K(n, \eta)$, hence $h_{K'}(g) \leq h_K(f)$. Consequently $h_K(f) = h_{K'}(g)$. Since we take the supremum over all compact sets and since p is surjective, we obtain $h_X(f) = h_Y(g)$. $\square$

We add one additional fact.

PROPOSITION 10.4. *If X is compact and $f\colon X \to X$ is a homeomorphism, then $h(f^n) = |n| h(f)$.*

For a proof, see [AKM65] or [Bow71].

10.2 THE FUNDAMENTAL GROUP AND ENTROPY

Given a finitely generated group G and a finite set of generators $\mathcal{G} = \{g_1, \ldots, g_r\}$ of G, we define the *length* of an element g of G by $L_{\mathcal{G}}(g) = $ minimum length of a word in the g_i's and the g_i^{-1}'s representing the element g.

It is easy to see that if $\mathcal{G}' = \{g_1', \ldots, g_s'\}$ is another set of generators, then

$$L_{\mathcal{G}}(g) \leq (\max L_{\mathcal{G}}(g_i')) L_{\mathcal{G}'}(g).$$

If $A\colon G \to G$ is an endomorphism, let

$$\gamma_A = \sup_{g \in G} \limsup \frac{1}{n} \log L_{\mathcal{G}}(A^n g) = \sup_{g_i \in \mathcal{G}} \limsup \frac{1}{n} \log L_{\mathcal{G}}(A^n g_i).$$

So γ_A is finite and by the inequality given above, γ_A does not depend on the set of generators.

PROPOSITION 10.5. *If $A\colon G \to G$ is an endomorphism and $g \in G$, define $gAg^{-1}\colon G \to G$ by $[gAg^{-1}](x) = gA(x)g^{-1}$. We have $\gamma_A = \gamma_{gAg^{-1}}$.*

Caution: $(gAg^{-1})^n \neq gA^n g^{-1}$.

First, we need a lemma.

LEMMA 10.6. *Let $(a_n)_{n \geq 1}$ and $(b_n)_{n \geq 1}$ be two sequences with a_n and b_n nonnegative, and let k be positive. We have*

(i) $\limsup \frac{1}{n} \log(a_n + b_n) = \max(\limsup \frac{1}{n} \log a_n, \limsup \frac{1}{n} \log b_n)$;

(ii) $\limsup \frac{1}{n} \log k a_n = \limsup \frac{1}{n} \log a_n$;

(iii) $\limsup \frac{1}{n} \log a_n \leq \limsup \frac{1}{n} \log(a_1 + \cdots + a_n)$

$$\leq \max(0, \limsup \frac{1}{n} \log a_n).$$

Proof. Set $a = \limsup \frac{1}{n} \log a_n$ and $b = \limsup \frac{1}{n} \log b_n$.

(i) The inequality $\max(a, b) \leq \limsup \frac{1}{n} \log(a_n + b_n)$ is clear.

If $c > \max(a, b)$, then we can find $n_0 \geq 1$ such that $n \geq n_0$ implies $a_n \leq e^{nc}$ and $b_n \leq e^{nc}$. We obtain for $n \geq n_0$

$$\frac{1}{n} \log(a_n + b_n) \leq \frac{1}{n} \log(2e^{nc}).$$

Hence $\limsup \frac{1}{n} \log(a_n + b_n) \leq \limsup \frac{1}{n} \log(2e^{nc}) = c$.

(ii) is clear.

(iii) The inequality $a \leq \limsup \frac{1}{n} \log(a_1 + \cdots + a_n)$ is clear.

Suppose $c > \max(0, a)$. We can find then $n_0 \geq 1$ such that $a_n \leq e^{nc}$ for $n \geq n_0$. We have for $n \geq n_0$

$$a_1 + \cdots + a_n \leq \sum_{i=1}^{n_0 - 1} a_i + \frac{e^{(n+1-n_0)c} - 1}{e^c - 1} e^{n_0 c}.$$

It follows clearly that $\limsup_n \frac{1}{n} \log(a_1 + \cdots + a_n) \leq c$. $\qquad\qquad\square$

Proof of Proposition 10.5. If $x \in G$, we have

$$(gAg^{-1})^n(x) = gA(g) \cdots A^{n-1}(g)A^n(x)A^{n-1}(g^{-1}) \cdots A(g^{-1})g^{-1}.$$

Suppose first that $A^{n_0}(g) = e$ for some n_0, then it is clear that by Lemma 10.6 *(i)*

$$\limsup \frac{1}{n} \log L_{\mathcal{G}}[(gAg^{-1})^n(x)] \leq \limsup \frac{1}{n} \log L_{\mathcal{G}}(A^n(x)).$$

If $A^n(g) \neq e$ for each $n \geq 1$, we have $L_{\mathcal{G}}(A^n(g)) \geq 1$ for each $n \geq 1$; hence $\limsup \frac{1}{n} \log L_{\mathcal{G}}(A^n(g)) \geq 0$. By Lemma 10.6 *(i)* and *(iii)*, we obtain

$$\limsup \frac{1}{n} \log L_{\mathcal{G}}[(gAg^{-1})^n(x)]$$
$$\leq \max(\limsup \frac{1}{n} \log L_{\mathcal{G}}((A^n(g)), \limsup \frac{1}{n} \log L_{\mathcal{G}}(A^n(x))).$$

This gives us $\gamma_{gAg^{-1}} \leq \gamma_A$, and by symmetry, we have $\gamma_{gAg^{-1}} = \gamma_A$. $\qquad\square$

For a compact, connected, differentiable manifold, we interpret $\pi_1(M)$ as the group of covering transformations of the universal covering space $\widetilde{M}$ of M. If $f\colon M \to M$ is continuous, then there is a lifting $\tilde{f}\colon \widetilde{M} \to \widetilde{M}$. If $\tilde{f}_1$ and $\tilde{f}_2$ are both liftings of f, then $\tilde{f}_1 = \theta \tilde{f}_2$ for some covering transformation θ. A given lifting $\tilde{f}_1$ determines an endomorphism $\tilde{f}_{1\#}$ of $\pi_1(M)$ by the formula

$$\tilde{f}_1 \alpha = \tilde{f}_{1\#}(\alpha)\tilde{f}_1$$

for any covering transformation α. If $\tilde{f}_1$ and $\tilde{f}_2$ are two liftings of f, then $\tilde{f}_1 = \theta \tilde{f}_2$ for some covering transformation θ and

$$\tilde{f}_1 \alpha = \theta \tilde{f}_2 \alpha = \theta \tilde{f}_{2\#}(\alpha)\tilde{f}_2 = \theta \tilde{f}_{2\#}(\alpha)\theta^{-1}\tilde{f}_1,$$

so $\tilde{f}_{1\#} = \theta \tilde{f}_{2\#}\theta^{-1}$ and $\gamma_{\tilde{f}_{1\#}} = \gamma_{\tilde{f}_{2\#}}$. Thus, we may define

$$\gamma_{f_\#} = \gamma_{\tilde{f}_\#}$$

for any lifting $\tilde{f}\colon \widetilde{M} \to \widetilde{M}$ of f. If f has a fixed point $m_0 \in M$, then there is also a map $f_\#\colon \pi_1(M, m_0) \to \pi_1(M, m_0)$. The group $\pi_1(M, m_0)$ is isomorphic to the group of covering transformations of $\widetilde{M}$ and f may be lifted to $\tilde{f}$ such that $\tilde{f}_\#\colon \pi_1(M) \to \pi_1(M)$ is identified with $f_\#\colon \pi_1(M, m_0) \to \pi_1(M, m_0)$ by this isomorphism. Thus $\gamma_{f_\#}$ makes coherent sense in the case that f has a fixed point as well.

We suppose now that M has a Riemannian metric and we endow $\widetilde{M}$ with a Riemannian metric by lifting the metric on M via the covering map $p\colon \widetilde{M} \to M$. The map p is then a metric covering and the covering transformations are isometries. We have the following lemma due to Milnor [Mil68].

LEMMA 10.7. *Fix $x_0 \in \widetilde{M}$. There exist two constants $c_1, c_2 > 0$ such that for each $g \in \pi_1(M)$, we have*

$$c_1 L_{\mathcal{G}}(g) \leq d(x_0, g x_0) \leq c_2 L_{\mathcal{G}}(g).$$

Proof [Mil68]. Let $\delta = \operatorname{diam}(M)$, and define $N \subset \widetilde{M}$ by

$$N = \{x \in \widetilde{M} \mid d(x, x_0) \leq \delta\}.$$

We have $p(N) = M$. Note that $\{gN\}_{g \in \pi_1(M)}$ is a locally finite covering of $\widetilde{M}$ by compact sets. Choose as a finite set of generators

$$\mathcal{G} = \{g \in \pi_1(M) \mid gN \cap N \neq \emptyset\}$$

and notice that $g \in \mathcal{G} \leftrightarrow g^{-1} \in \mathcal{G}$. Suppose $L_{\mathcal{G}}(g) = n$, then we can write $g = g_1 \cdots g_n$, with $g_i N \cap N \neq \emptyset$. It is easy to see then that $d(x_0, g x_0) \leq 2\delta n$. Hence, we obtain

$$d(x_0, g x_0) \leq 2\delta L_{\mathcal{G}}(g).$$

Now, set $\nu = \min\{d(N, gN) \mid N \cap gN = \emptyset\}$; by compactness $\nu > 0$. Let k be the minimal integer such that $d(x_0, g x_0) < k\nu$. Along the minimizing geodesic from x_0 to $g x_0$, take $k + 1$ points $y_0 = x_0, y_1, \ldots, y_{k-1}, y_k = g x_0$ such that $d(y_i, y_{i+1}) < \nu$ for $i = 0, \ldots, k - 1$. Then, for $1 \leq i \leq k - 1$, choose $y_i' \in N$ and $g_i \in G$ such that $y_i = g_i y_i'$ and set $g_0 = e$ and $g_k = g$. We have $d(g_i y_i', g_{i+1} y_{i+1}') < \nu$, hence $g_i^{-1} g_{i+1} \in \mathcal{G}$. From $g = (g_0^{-1} g_1) \cdots (g_{k-1}^{-1} g_k)$, we obtain $L_{\mathcal{G}}(g) < k$.

Since k is minimal, we have

$$L_{\mathcal{G}}(g) \leq \frac{1}{\nu} d(x_0, g x_0) + 1 \leq \left(\frac{1}{\nu} + \frac{1}{\mu}\right) d(x_0, g x_0)$$

where $\mu = \min\{d(x_0, g x_0) \mid g \neq e, g \in \pi_1(M)\}$. $\qquad\square$

Consider now $f : M \to M$ and let $\tilde{f} : \widetilde{M} \to \widetilde{M}$ be a lifting of f. Applying the lemma above, we obtain, for each $x_0 \in \widetilde{M}$:

$$\gamma_{f_\#} = \max_{g \in \pi_1(M)} \limsup \frac{1}{n} \log d(x_0, \tilde{f}_\#^n(g) x_0).$$

We next prove the following lemma.

LEMMA 10.8. *Given $x, y \in \widetilde{M}$, we have*

$$\limsup \frac{1}{n} \log d(\tilde{f}^n(x), \tilde{f}^n(y)) \leq h(f).$$

Proof. Choose an arc α from x to y. If $y_1, \ldots, y_\ell \in \alpha$ is $(n + 1, \epsilon)$-spanning for α and $\tilde{f}$, then

$$\tilde{f}^n(\alpha) \subset \bigcup_{i=1}^{\ell} B(\tilde{f}^n(y_i), \epsilon).$$

Since $\tilde{f}^n(\alpha)$ is connected, this implies $\operatorname{diam}(\tilde{f}^n(\alpha)) < 2\epsilon\ell$. Hence

$$d(\tilde{f}^n(x), \tilde{f}^n(y)) \leq 2\epsilon\ell.$$

By taking ℓ to be minimal, we obtain

$$d(\tilde{f}^n(x), \tilde{f}^n(y)) \leq 2\epsilon r_\alpha(n+1, \epsilon).$$

From this, we get

$$
\begin{aligned}
\limsup_{n\to\infty} \frac{1}{n} \log d(\tilde{f}^n(x), \tilde{f}^n(y)) &\leq \limsup_{n\to\infty} \frac{1}{n} \log[2\epsilon r_\alpha(n+1, \epsilon)] = \bar{r}_\alpha(\epsilon) \\
&\leq h_\alpha(\tilde{f}) \leq h(\tilde{f}) = h(f).
\end{aligned}
$$

$\square$

We are now ready to prove the following.

THEOREM 10.9. *If $f\colon M \to M$ is a continuous map, then*

$$h(f) \geq \gamma_{f_\#}.$$

Proof. Since

$$\gamma_{f_\#} = \max_{g\in\pi_1(M)} [\limsup \frac{1}{n} \log d(x_0, \tilde{f}^n_\#(g)x_0)],$$

we have to prove that for each $g \in \pi_1(M)$,

$$\limsup \frac{1}{n} \log d(x_0, \tilde{f}^n_\#(g)x_0) \leq h(f).$$

We have

$$
\begin{aligned}
d(x_0, \tilde{f}^n_\#(g)x_0) &\leq d(x_0, \tilde{f}^n(x_0)) + \\
&\quad d(\tilde{f}^n(x_0), \tilde{f}^n_\#(g)\tilde{f}^n(x_0)) + \\
&\quad d(\tilde{f}^n_\#(g)\tilde{f}^n(x_0), \tilde{f}^n_\#(g)x_0).
\end{aligned}
$$

Since $\tilde{f}^n_\#(g)\tilde{f}^n = \tilde{f}^n g$, and the covering transformations are isometries, we obtain

$$d(x_0, \tilde{f}^n_\#(g)x_0) \leq 2d(x_0, \tilde{f}^n(x_0)) + d(\tilde{f}^n(x_0), \tilde{f}^n g(x_0)).$$

Remark also that

$$
\begin{aligned}
d(x_0, \tilde{f}^n(x_0)) &\leq d(x_0, \tilde{f}(x_0)) + d(\tilde{f}(x_0), \tilde{f}^2(x_0)) + \cdots + \\
&\quad d(\tilde{f}^{n-1}(x_0), \tilde{f}^n(x_0)).
\end{aligned}
$$

By applying Lemma 10.8 and Lemma 10.6 (together with the fact that $h(f) \geq 0$), we obtain

$$\limsup \frac{1}{n} \log d(x_0, \tilde{f}^n_\#(g)x_0) \leq h(f).$$

$\square$

The proof of the following lemma is straightforward.

LEMMA 10.10. *If G_1 and G_2 are finitely generated groups, and if $A\colon G_1 \to G_1$, $B\colon G_2 \to G_2$ and $p\colon G_1 \to G_2$ are homomorphisms with p surjective and $pA = Bp$,*

$$
\begin{array}{ccc}
G_1 & \xrightarrow{\ p\ } & G_2 \longrightarrow 0 \\
{\scriptstyle A}\big\downarrow & & {\scriptstyle B}\big\downarrow \\
G_1 & \xrightarrow{\ p\ } & G_2 \longrightarrow 0
\end{array}
$$

then $\gamma_A \geq \gamma_B$.

Applying this lemma to the fundamental group of M modulo the commutator subgroup, we have

$$
\begin{array}{ccc}
\pi_1(M) & \xrightarrow{\ p\ } & H_1(M) \longrightarrow 0 \\
{\scriptstyle f_\#}\big\downarrow & & {\scriptstyle f_{1*}}\big\downarrow \\
\pi_1(M) & \xrightarrow{\ p\ } & H_1(M) \longrightarrow 0
\end{array}
$$

so we obtain Manning's theorem [Man75].

THEOREM 10.11. *If $f\colon M \to M$ is continuous, then*

$$
h(f) \geq \gamma_{f_{1*}} = \max \log \lambda,
$$

where λ ranges over the eigenvalues of f_{1*}.

Remark 1. For $\alpha \in \pi_1(M, m_0)$, we denote by $[\alpha]$ the class of loops freely homotopic to α. If M has a Riemannian metric, let $\ell([\alpha])$ be the minimum length of a (smooth) loop in this class. If $f\colon M \to M$ is continuous, $f[\alpha]$ is clearly well-defined as a free homotopy class of loops. Let

$$
G_f([\alpha]) - \limsup_n \frac{1}{n} \log[\ell(f^n[\alpha])]
$$

and let $G_f = \sup_\alpha G_f([\alpha])$.

It is not difficult to see that $G_f \leq \gamma_{f_\#}$. In fact, we have $\ell(f^n[\alpha]) \leq d(x_0, \tilde{f}^n_\#(\alpha)(x_0))$, since the minimizing geodesic from x_0 to $\tilde{f}^n_\#(\alpha)(x_0)$ has an image in M that represents $f^n[\alpha]$.

Remark 2. It occurred to various people that Manning's theorem is a theorem about fundamental groups. Among these are Bowen, Gromov, and Shub. Manning's proof can be adapted. The proof above is more like Gromov [Gro00] or Bowen [Bow71], but we take responsibility for any error. At first, we assumed that f had a periodic point or we worked with G_f. After reading Bowen's proof [Bow78], we eliminated the necessity for a periodic point.

Remark 3. If $x \in M$ and ρ is a path joining x to $f(x)$, we call $\rho_\#$ the homomorphism $\pi_1(M, f(x)) \to \pi_1(M, x)$. Since $f_\# \colon \pi_1(M, x) \to \pi_1(M, f(x))$, the composition

$$\rho_\# f_\# \colon [\gamma] \mapsto [\rho^{-1}\gamma\rho]$$

is a homomorphism of $\pi_1(M, x)$ into itself. This homomorphism can be identified with $\tilde{f}_\#$ for a lifting $\tilde{f}$ of f. Thus our result is the same as Bowen's [Bow78].

10.3 SUBSHIFTS OF FINITE TYPE

Let $A = (a_{ij})$ be a $k \times k$ matrix such that $a_{ij} = 0$ or 1, for $1 \leq i, j \leq k$, that is, A is a "0–1 matrix." Such a matrix A determines a subshift of finite type as follows. Let $S_k = \{1, \dots, k\}$ and let

$$\Sigma(k) = \prod_{i=-\infty}^{i=\infty} S_k^i,$$

where $S_k^i = S_k$ for each $i \in \mathbb{Z}$. We endow S_k with the discrete topology and $\Sigma(k)$ with the product topology. The subset $\Sigma_A \subset \Sigma(k)$ is the closed subset consisting of those bi-infinite sequences $\underline{b} = (b_n)_{n \in \mathbb{Z}}$ such that $a_{b_i b_{i+1}} = 1$ for all $i \in \mathbb{Z}$.

Pictorially, we imagine k boxes

$$\boxed{1}\,\boxed{2}\;\cdots\;\boxed{k}$$

and a point that at discrete "time n" can be in any one of the boxes. The bi-infinite sequences represent all possible histories of points. If we add the restriction that a point may move from box i to box j if and only if $a_{ij} = 1$, then the set of all possible histories is precisely Σ_A.

The *shift* $\sigma_A \colon \Sigma_A \to \Sigma_A$ is defined by $\sigma_A[(b_n)_{n \in \mathbb{Z}}] = (b'_n)_{n \in \mathbb{Z}}$ where $b'_n = b_{n+1}$ for each $n \in \mathbb{Z}$. Clearly, σ_A is continuous. Let $C_i \subset \Sigma(k)$ be defined by $C_i = \{\underline{x} \in \Sigma(k) \mid x_0 = i\}$. Let $D_i = C_i \cap \Sigma_A$, then $\mathcal{D} = \{D_1, \dots, D_k\}$ is an open cover of Σ_A by pairwise disjoint elements. For any $k \times k$ matrix $B = (b_{ij})$, we define the norm $||B||$ of B by

$$||B|| = \sum_{i,j=1}^{k} |b_{ij}|.$$

It is easy to see that

$$N_n(\sigma_A, \mathcal{D}) = \mathrm{card}(\mathcal{D} \vee \cdots \vee \sigma_A^{-n+1}\mathcal{D}) \leq ||A^{n-1}||$$

because the integer $a_{ij}^{(n)}$ is equal to the number of sequences $(i_0, \dots, i_n)$ with $i_\ell \in \{1, \dots, k\}$, $i_0 = i$, $i_n = j$, and $a_{i_\ell i_{\ell+1}} = 1$. So

$$\limsup \frac{1}{n} \log(N_n(\sigma_A, \mathcal{D})) \;\leq\; \limsup \frac{1}{n} \log ||A^{n-1}||$$

$$= \; \limsup \log ||A^n||^{1/n}.$$

This latter number is recognizable as $\log(\text{spectral radius } A)$ or $\log \lambda$, where λ is the largest modulus of an eigenvalue of A. In fact, we have:

PROPOSITION 10.12. *For any subshift of finite type* $\sigma_A \colon \Sigma_A \to \Sigma_A$, *we have* $h(\sigma_A) = \log \lambda$, *where* λ *is the spectral radius of* A.

Proof. We begin by noticing that each open cover $\mathcal{U}$ of Σ_A is refined by a cover of the form $\bigvee_{i=-\ell}^{\ell} \sigma_A^{-i} \mathcal{D}$. This implies, with the notations of Section 10.1,

$$
\begin{aligned}
N_{n+1}(\sigma_A, \mathcal{U}) &\leq \operatorname{card}\left(\bigvee_{j=-\ell}^{j=n+\ell} \sigma_A^{-j} \mathcal{D} \right) \\
&= \operatorname{card}\left(\bigvee_{j=0}^{j=n+2\ell} \sigma_A^{-j} \mathcal{D} \right) \\
&= N_{n+2\ell+1}(\sigma_A, \mathcal{D}).
\end{aligned}
$$

Hence, we obtain $h(\sigma_A, \mathcal{U}) \leq h(\sigma_A, \mathcal{D})$.

This shows that $h(\sigma_A) = h(\sigma_A, \mathcal{D})$.

We now compute $h(\sigma_A, \mathcal{D})$. We distinguish two cases.

First case: Each state $i = 1, \dots, k$ *occurs.* This means that $D_i \neq \emptyset$ for each $D_i \in \mathcal{D}$. It is not difficult to show by induction that we have in fact

$$
N_{n+1}(\sigma_A, \mathcal{D}) = \operatorname{card}(\mathcal{D} \vee \cdots \vee \sigma_A^{-n} \mathcal{D}) = \|A^n\|.
$$

This proves the proposition in this case, as we saw above.

Second case: Some states do not occur. One can see that a state i occurs if and only if for each $n \geq 0$, we have

$$
\sum_{j=1}^{k} a_{ij}^{(n)} > 0 \quad \text{and} \quad \sum_{j=1}^{k} a_{ji}^{(n)} > 0
$$

where $A^n = \left(a_{ij}^{(n)} \right)$.

Notice that if $\sum_{j=1}^{k} a_{ij}^{(n_0)} = 0$, then $\sum_{j=1}^{k} a_{ij}^{(n)} = 0$ for all $n \geq n_0$. This is because each $a_{\ell m}$ is nonnegative.

Now, we partition $\{1,\dots,k\}$ into three subsets X, Y, Z, where

$$X = \{i \mid \forall n \geq 0 \text{ we have } \sum_{j=1}^{k} a_{ij}^{(n)} > 0 \text{ and } \sum_{j=1}^{k} a_{ji}^{(n)} > 0\},$$

$$Y = \{i \mid \exists n > 0 \text{ so that } \sum_{j=1}^{k} a_{ij}^{(n)} = 0\}$$

$$= \{i \mid \text{for } n \text{ large we have } \sum_{j=1}^{k} a_{ij}^{(n)} = 0, \}$$

$$Z = \{1,\dots,k\} - (X \cup Y).$$

We have

$$Z \subset \{i \mid \text{for } n \text{ large } \sum_{j=1}^{k} a_{ji}^{(n)} = 0\}.$$

By performing a permutation of $\{1,\dots,k\}$, we can suppose that we have the following situation:

$$\{\underbrace{1,\dots,t}_{X}, \underbrace{t+1,\dots,s}_{Y}, \underbrace{s+1,\dots,k}_{Z}\}.$$

If B is a $k \times k$ matrix, we write

$$B = \begin{pmatrix} B_{XX} & B_{XY} & B_{XZ} \\ B_{YX} & B_{YY} & B_{YZ} \\ B_{ZX} & B_{ZY} & B_{ZZ} \end{pmatrix}$$

where B_{KL} corresponds to the subblock of B having row indices in K and column indices in L.

It is easy to show that

$$N_{n+1}(\sigma_A, \mathcal{D}) = \mathrm{card}(\mathcal{D} \vee \cdots \vee \sigma_A^{-n}\mathcal{D}) = \|A_{X,X}^n\|.$$

On the other hand, by the definition of Y and Z, for n large, A^n has the form

$$A^n = \begin{bmatrix} (A^n)_{X,X} & (A^n)_{X,Y} & 0 \\ 0 & 0 & 0 \\ (A^n)_{Z,X} & (A^n)_{Z,Y} & 0 \end{bmatrix}$$

This implies that for n large, A^n and $(A^n)_{X,X}$ have the same nonzero eigenvalues. In particular

$$\log(\text{spectral radius } A_{X,X}^n) = n \log \lambda.$$

Also note that, for n large and $k \geq 1$, we have

$$(A^{kn})_{X,X} = [(A^n)_{X,X}]^k.$$

This gives us, for n large,

$$\limsup_{k\to\infty} \frac{\log N_{kn+1}(\sigma_A, \mathcal{D})}{kn+1} = \limsup_{k\to\infty} \frac{\|[(A^n)_{X,X}]^k\|}{kn+1} = \log \lambda.$$

This implies that

$$\log \lambda \leq h(\sigma_A, \mathcal{D}) = \limsup_{n\to\infty} \frac{1}{n} \log N_n(\sigma_A, \mathcal{D}).$$

As we showed the reverse inequality, we have

$$\log \lambda = h(\sigma_A, \mathcal{D}) = h(\sigma_A).$$

$\square$

10.4   THE ENTROPY OF PSEUDO-ANOSOV DIFFEOMORPHISMS

Now we suppose that we have a compact, connected 2-manifold M without boundary and with genus at least 2, and a pseudo-Anosov diffeomorphism $f\colon M \to M$. Hence there exists a pair $(\mathcal{F}^u, \mu^u)$ and $(\mathcal{F}^s, \mu^s)$ of transverse measured foliations with (the same) singularities such that $f(\mathcal{F}^s, \mu^s) = (\mathcal{F}^s, \frac{1}{\lambda}\mu^s)$ and $f(\mathcal{F}^u, \mu^u) = (\mathcal{F}^u, \lambda\mu^u)$ where $\lambda > 1$. This means, in particular, that f preserves the two foliations $\mathcal{F}^s$ and $\mathcal{F}^u$; it contracts the leaves of $\mathcal{F}^s$ by $1/\lambda$ and it expands the leaves of $\mathcal{F}^u$ by λ.

Let us recall that for any nontrivial simple closed curve α we have $\log \lambda = G_f([\alpha])$ (see Proposition 9.21), hence we get $\log \lambda \leq G_f$. [For the definition of G_f, see the end of Section 10.2.]

PROPOSITION 10.13. *If $f\colon M \to M$ is pseudo-Anosov, then $h(f) = \gamma_{f_\#}$. So in particular, f has the minimal entropy of any element of its homotopy class. Moreover, $h(f) = \log \lambda$ where λ is the expanding factor of f.*

Proof. Since $G_f \geq \log \lambda$, it suffices to show that $h(f) \leq \log \lambda$ for a pseudo-Anosov diffeomorphism f. To do this, we find a subshift of finite type $\sigma_A\colon \Sigma_A \to \Sigma_A$ and a surjective continuous map $\Sigma_A \to M$ such that the diagram

$$\begin{array}{ccc} \Sigma_A & \xrightarrow{\ \sigma_A\ } & \Sigma_A \\ \theta\Big\downarrow & & \Big\downarrow\theta \\ M & \xrightarrow{\ f\ } & M \end{array}$$

commutes, and $\log(\text{spectral radius } A) = h(\sigma_A) = \log \lambda$ for this same λ. Thus, we will have

$$\log \lambda \leq G_f \leq \gamma_{f_\#} \leq h(f) \leq h(\sigma_A)$$

or

$$\log \lambda \le h(f) \le \log \lambda.$$

$\square$

In the following, we construct A and θ via Markov partitions. First, some definitions.

Birectangles. A subset R of M is called an $(\mathcal{F}^s, \mathcal{F}^u)$-*rectangle*, or *birectangle*, if there exists an immersion $\varphi \colon [0,1] \times [0,1] \to M$ whose image is R and such that:

- $\varphi|_{(0,1) \times (0,1)}$ is an embedding.

- $\forall\, t \in [0,1], \varphi(\{t\} \times [0,1])$ is included in a finite union of leaves and singularities of $\mathcal{F}^s$, and in fact in one leaf if $t \in (0,1)$.

- $\forall\, t \in [0,1], \varphi([0,1] \times \{t\})$ is included in a finite union of leaves and singularities of $\mathcal{F}^u$, and in fact in one leaf if $t \in (0,1)$.

We adopt the following notations:

$$
\begin{aligned}
\mathrm{int}(R) &= \varphi\big((0,1) \times (0,1)\big), \\
\partial^0_{\mathcal{F}^s} R &= \varphi(\{0\} \times [0,1]), \\
\partial^1_{\mathcal{F}^s} R &= \varphi(\{1\} \times [0,1]), \\
\partial_{\mathcal{F}^s} R &= \partial^0_{\mathcal{F}^s} R \;\cup\; \partial^1_{\mathcal{F}^s} R,
\end{aligned}
$$

and in the same way, we define $\partial^0_{\mathcal{F}^u} R$, $\partial^1_{\mathcal{F}^u} R$, $\partial_{\mathcal{F}^u} R$.

Note that $\mathrm{int}(R)$ is disjoint from $\partial_{\mathcal{F}^s} R \;\cup\; \partial_{\mathcal{F}^u} R$, because $\varphi|(0,1) \times (0,1)$ is an embedding.

We call a set of the form $\varphi(\{t\} \times [0,1])$ (resp. $\varphi([0,1] \times \{t\})$) an $\mathcal{F}^s$-fiber (resp. an $\mathcal{F}^u$-fiber) of R. We will call a birectangle *good* if φ is an embedding.

If R is good birectangle, a point x of R is contained in only one $\mathcal{F}^s$-fiber, which we will denote by $\mathcal{F}^s(x, R)$. In the same way, we define $\mathcal{F}^u(x, R)$.

Remark 1. If R is an $\mathcal{F}^u$-rectangle (see Exposé 9) and $\partial^0_\tau R$ and $\partial^1_\tau R$ are contained in a union of $\mathcal{F}^s$-leaves and singularities, it is easy to see that R is in fact a birectangle.

Remark 2. We used the word birectangle instead of rectangle, even though rectangle is the standard word in Markov partitions, because this word was already used in Exposé 9.

Remark 3. If R_1 and R_2 are birectangles and $R_1 \cap R_2 \ne \emptyset$, then it is a finite union of birectangles and possibly of some arcs contained in $(\partial_{\mathcal{F}^s} R_1 \cup \partial_{\mathcal{F}^u} R_1) \cap (\partial_{\mathcal{F}^s} R_2 \cup \partial_{\mathcal{F}^u} R_2)$. Moreover, the birectangles are the closures of the connected components of $\mathrm{int}(R_1) \cap \mathrm{int}(R_2)$.

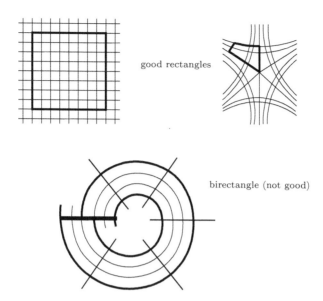

good rectangles

birectangle (not good)

If R is a birectangle, we define the *width* of R by

$$\mathcal{W}(R) = \max\{\mu^u(\mathcal{F}^s\text{-fiber}), \mu^s(\mathcal{F}^u\text{-fiber})\}.$$

LEMMA 10.14. *There exists $\epsilon > 0$ such that, if R is a birectangle with $\mathcal{W}(R) \leq \epsilon$, then it is a good rectangle.*

Sketch. If a birectangle is contained in a coordinate chart of the foliations, then it is automatically a good birectangle. The existence of ϵ follows from compactness. □

LEMMA 10.15. *There exists $\epsilon > 0$ such that if α (resp. β) is an arc contained in a finite union of leaves and singularities of $\mathcal{F}^s$ (resp. $\mathcal{F}^u$) with $\mu^u(\alpha) < \epsilon$ (resp. $\mu^s(\beta) < \epsilon$), then the intersection of α and β is at most one point.*

Markov partitions. A *Markov partition* for the pseudo-Anosov diffeomorphism $f: M \to M$ is a collection R of birectangles $\{R_1, \ldots, R_k\}$ such that

1. $\bigcup_{i=1}^{k} R_i = M$.

2. R_i is a good rectangle.

3. $\text{int}(R_i) \cap \text{int}(R_j) = \emptyset$ for $i \neq j$.

4. If $x \in \text{int}(R_i)$ and $f(x) \in \text{int}(R_j)$, then

$$f(\mathcal{F}^s(x, R_i)) \subset \mathcal{F}^s(f(x), R_j)$$

and
$$f^{-1}(\mathcal{F}^u(f(x), R_j)) \subset \mathcal{F}^u(x, R_i).$$

5. If $x \in \mathrm{int}(R_i)$ and $f(x) \in \mathrm{int}(R_j)$, then
$$f(\mathcal{F}^u(x, R_i)) \cap R_j = \mathcal{F}^u(f(x), R_j)$$

and
$$f^{-1}(\mathcal{F}^s(x, R_j)) \cap R_i = \mathcal{F}^s(x, R_i).$$

This means that $f(R_i)$ goes across R_j just one time.

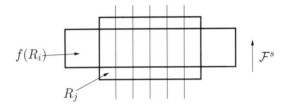

We will show in the next section how to construct a Markov partition for a pseudo-Anosov diffeomorphism.

Given a Markov partition $\mathcal{R} = \{R_1, \ldots, R_k\}$, we construct the subshift of finite type Σ_A and the map $h \colon \Sigma_A \to M$ as follows. Let A be the $k \times k$ matrix defined by $a_{ij} = 1$ if $f(\mathrm{int}(R_i)) \cap \mathrm{int}(R_j) \neq \emptyset$, and $a_{ij} = 0$ otherwise. If $\underline{b} \in \Sigma_A$, then
$$\bigcap_{i \in \mathbb{Z}} f^{-i}(R_{b_i})$$

is nonempty—it is, in fact, a single point. This will follow from Lemma 10.16.

LEMMA 10.16. (i) Suppose $a_{ij} = 1$, then $f(R_i) \cap R_j$ is a nonempty (good) birectangle which is a union of $\mathcal{F}^u$-fibers of R_j.

(ii) Suppose moreover that C is a birectangle contained in R_i which is a union of $\mathcal{F}^u$-fibers of R_i. Then $f(C) \cap R_j$ is a nonempty birectangle which is a union of $\mathcal{F}^u$-fibers of R_j.

(iii) Given $\underline{b} \in \Sigma_A$, for each $n \in \mathbb{N}$,
$$\bigcap_{i=-n}^{n} f^{-i}(R_{b_i})$$

is a nonempty birectangle. Moreover, we have
$$\mathcal{W}\left(\bigcap_{i=-n}^{n} f^{-i}(R_{b_i})\right) \leq \lambda^{-n} \max\{\mathcal{W}(R_1), \ldots, \mathcal{W}(R_k)\}.$$

Proof. Since $a_{ij} = 1$, we can find $x \in \text{int}(R_i) \cap f^{-1}(\text{int}(R_j))$. Then

$$f(\mathcal{F}^s(x, R_i)) \subset \mathcal{F}^s(f(x), R_j) \subset R_j.$$

Since each $\mathcal{F}^u$-fiber of R_i intersects $\mathcal{F}^s(x, R_i)$, we obtain that the image of each $\mathcal{F}^u$-fiber of R_i intersects R_j. Moreover, by condition 5 we have that $f[R_i - \partial_{\mathcal{F}^u} R_i] \cap R_j$ is a union of $\mathcal{F}^u$-fibers of R_j, hence

$$f(R_i) \cap R_j = \overline{f(R_i - \partial_{\mathcal{F}^u} R_i) \cap R_j}$$

is also a union of $\mathcal{F}^u$-fibers of R_j. This proves *(i)*. The proof of *(ii)* is the same.

To prove *(iii)*, remark first that it follows by induction on n using *(ii)* that each set of the form $f^n(R_{b_i}) \cap f^{n-1}(R_{b_{i+1}}) \cap \cdots \cap R_{b_{i+n}}$ is a nonempty birectangle which is a union of $\mathcal{F}^u$-fibers of $R_{b_{i+n}}$. In particular,

$$\bigcap_{i=-n}^{n} f^{-i}(R_{b_i})$$

is a nonempty birectangle in R_{b_0}. The estimate of the width is clear. □

By the lemma, if $\underline{b} \in \Sigma_A$, the set $\bigcap f^{-i}(R_{b_i})$ is the intersection of a decreasing sequence of nonempty compact sets, namely, the sets

$$\bigcap_{i=-n}^{n} f^{-i}(R_{b_i})$$

for $n \in \mathbb{N}$. Hence $\bigcap f^{-i}(R_{b_i})$ is nonempty. It is reduced to one point because

$$\mathcal{W}\left(\bigcap_{i=-n}^{n} f^{-i}(R_{b_i})\right)$$

tends to zero as n goes to infinity

The map $\theta \colon \Sigma_A \to M$ given by

$$\theta(\underline{b}) = \bigcap_{i \in \mathbb{Z}} f^{-i}(R_{b_i})$$

is well-defined, and it is easy to see that it is continuous and that the following diagram commutes:

$$
\begin{array}{ccc}
\Sigma_A & \xrightarrow{\sigma_A} & \Sigma_A \\
\downarrow{\theta} & & \downarrow{\theta} \\
M & \xrightarrow{f} & M
\end{array}
$$

We show now that θ is surjective. First remark that, for each $i = 1, \ldots, k$, the closure of $\text{int}(R_i)$ is R_i. Hence

$$V = \bigcup_{i=1}^{k} \text{int}(R_i)$$

is a dense open set. By the Baire category theorem,

$$U = \bigcap_{i \in \mathbb{Z}} f^{-i}(V)$$

is dense in M. If $x \in U$, then for each $n \in \mathbb{Z}$, the point $f^n(x)$ is in a unique $\text{int}(R_{b_n})$ and $\underline{b} = \{b_n\}_{n \in \mathbb{Z}}$ is an element of Σ_A. It is clear that $\theta(\underline{b}) = x$. Thus $\theta(\Sigma_A) \supset U$. As Σ_A is compact and f continuous, we have $\theta(\Sigma_A) = M$.

Up to now, we have obtained that

$$\log \lambda \leq G_f \leq \gamma_{f_\#} \leq h(f) \leq h(\sigma_A) = \log(\text{spectral radius of } A).$$

All that remains is to show that

$$(\text{spectral radius of } A) = \lambda.$$

To see this, we do the following. Set $y_i = \mu^u(\mathcal{F}^s\text{-fiber of } R_i)$; it is clear that this quantity is independent of the $\mathcal{F}^s$-fiber of R_i and also $y_i > 0$.

We have the following trivial equality:

$$y_j = \sum_{i=1}^{k} \frac{y_i}{\lambda} a_{ij},$$

which gives

$$\lambda y_j = \sum_{i=1}^{k} y_i a_{ij}$$

(in particular λ is an eigenvalue of A). Hence, we obtain

$$\lambda y_j \geq \left(\sum_{i=1}^{k} a_{ij} \right) \min_i y_i.$$

This gives

$$\lambda \left(\sum_j y_j \right) \geq ||A|| \min_i y_i$$

where $|| \; ||$ is the norm introduced in Section 10.3.

In the same way, we obtain for each $n \geq 2$

$$\lambda^n \left(\sum_j y_j \right) \geq ||A^n|| \min_i y_i.$$

Hence

$$\lambda \geq ||A^n||^{1/n} \Big(\frac{\min(y_1, \ldots, y_k)}{\sum_j y_j} \Big)^{1/n}.$$

Since $\min(y_1, \ldots, y_k) > 0$,

$$\lim_{n \to \infty} \Big(\frac{\min(y_1, \ldots, y_k)}{\sum_j y_j} \Big)^{1/n} = 1.$$

We thus obtain

$$\lambda \geq \lim_{n \to \infty} ||A^n||^{1/n} = \text{spectral radius of } A.$$

Since λ is an eigenvalue of A, we obtain

$$\lambda = \text{spectral radius of } A.$$

10.5 CONSTRUCTING MARKOV PARTITIONS FOR PSEUDO-ANOSOV DIFFEOMORPHISMS

In this section, we still consider $f\colon M \to M$ a pseudo-Anosov diffeomorphism and we keep the notations of the last section. We sketch the proof of the following proposition.

PROPOSITION 10.17. *A pseudo-Anosov diffeomorphism has a Markov partition.*

Proof. Using the methods given in Section 9.5, it is easy, starting with a family of transversals to $\mathcal{F}^u$ contained in $\mathcal{F}^s$-leaves and singularities, to construct a family $\mathcal{R}$ of $\mathcal{F}^u$-rectangles $R_1, \ldots, R_\ell$, such that

(i) $\displaystyle\bigcup_{i=1}^{\ell} R_i = M;$

(ii) $\text{int}(R_i) \cap \text{int}(R_j) = \emptyset$ for $i \neq j;$

(iii) $f^{-1}\Big(\displaystyle\bigcup_{i=1}^{\ell} \partial_{\mathcal{F}^u} R_i \Big) \subset \displaystyle\bigcup_{i=1}^{\ell} \partial_{\mathcal{F}^u} R_i, \ f\Big(\displaystyle\bigcup_{i=1}^{\ell} \partial_{\mathcal{F}^s} R_i \Big) \subset \displaystyle\bigcup_{i=1}^{\ell} \partial_{\mathcal{F}^s} R_i.$

By the remark following the definition of birectangles, the R_i's are birectangles since the system of transversals is contained in $\mathcal{F}^s$-leaves and singularities.

We define for each n a family of birectangles $\{\mathcal{R}_n\}$ in the following way: the birectangles of $\{\mathcal{R}_n\}$ will be the closures of the connected components of the nonempty open sets contained in

$$\bigvee_{i=-n}^{n} f^i(\text{int}(R)) = \left\{ \bigcap_{i=-n}^{n} f^i(\text{int}(R_{a_i})) \mid R_{a_i} \in \mathcal{R} \right\}.$$

It is easy to see that $\mathcal{R}_n$ still satisfies the properties *(i)*, *(ii)*, and *(iii)* given above. Moreover, if $R \in \mathcal{R}_n$, we have

$$\mathcal{W}(R) \le \lambda^{-n} \max\{\mathcal{W}(R_i) \mid R_i \in \mathcal{R}\}.$$

In particular, by Lemma 10.14, for n sufficiently large, each birectangle in $\mathcal{R}_n$ is a good one.

We assert that for n sufficiently large $\mathcal{R}_n$ is a Markov partition. All that remains is to verify properties 4 and 5 of a Markov partition. It is an easy exercise to show that property 4 is a consequence of property *(iii)* given above (see Lemma 9.10). By Lemma 10.15, if n is sufficiently large and $R, R' \in \mathcal{R}_n$, then if $x \in R$, $f(\mathcal{F}^u(x, R))$ intersects each $\mathcal{F}^s$-fiber of R' in at most one point. Property 5 follows easily from the combination of this fact with property 4. $\square$

Example of a Markov partition on T^2. Let $A \colon T^2 \to T^2$ be the linear map defined by

$$A = \begin{pmatrix} 2 & 1 \\ 1 & 1 \end{pmatrix}.$$

Here $T^2 = \mathbb{R}^2/\mathbb{Z}^2$ and A acts on $\mathbb{R}^2$ preserving $\mathbb{Z}^2$; thus A defines a map of T^2. The translates of the eigenspaces of A foliate T^2. The map A on T^2 is Anosov. The foliation of T^2 corresponding to the eigenvalue $\frac{3+\sqrt{5}}{2}$ is expanded, while the foliation corresponding to $\frac{3-\sqrt{5}}{2}$ is contracted.

We draw in Figure 10.1 a fundamental domain with eigenspaces approximately drawn in.

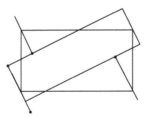

Figure 10.1.

The endpoints of the short stable manifold are on the unstable manifolds after equivalences have been made. Filling in to maximal rectangles gives us the picture in Figure 10.2.

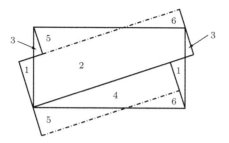

Figure 10.2.

The dashed line is the extension of the unstable manifold. Identified pieces are numbered similarly. One rectangle is given by 1,2,3,6 and the other by 4,5. This partition in two rectangles gives a Markov partition by taking intersections with direct and inverse images.

The construction of the Markov partition of a pseudo-Anosov diffeomorphism $f: M \to M$ that preserves orientation and fixes the prongs of $\mathcal{F}^s$ and $\mathcal{F}^u$ is the same as in the example above. We sketch here the argument, hoping that it will aid the reader to understand the general case.

Since the unstable prongs are dense, we may pick small stable prongs whose endpoints lie on unstable prongs. Roughly, the picture is

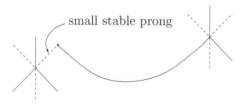

We may extend these curves to maximal birectangles leaving the drawn curves as boundaries. By density of the leaves, every leaf crosses a small stable prong, so the rectangles obtained this way cover M. The extension process requires that the unstable prongs be extended perhaps but the extension remains connected. Thus we have a partition by birectangles with boundaries the unions of connected segments lying on stable or unstable prongs. Consequently an unstable leaf entering the interior of a birectangle under f cannot end in the interior, because the stable boundary has been taken to the stable boundary, etc.

The only thing left is to make the partition sufficiently small. To do this, it is sufficient to take the birectangles obtained by intersections $f^{-n}(\mathcal{R}) \vee \cdots \vee \mathcal{R} \vee \cdots \vee f^{n}(\mathcal{R})$ for n sufficiently large.

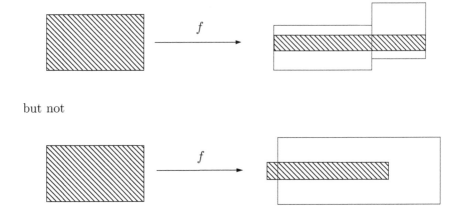

but not

10.6 PSEUDO-ANOSOV DIFFEOMORPHISMS ARE BERNOULLI

A pseudo-Anosov diffeomorphism $f\colon M \to M$ has a natural invariant probability measure μ that is given locally by the product of μ^s restricted to plaques of $\mathcal{F}^u$ with μ^u restricted to plaques of $\mathcal{F}^s$. The goal of this section is to sketch the proof of the following theorem.

THEOREM 10.18. *The dynamical system (M, f, μ) is isomorphic (in the measure theoretical sense) to a Bernoulli shift.*

Recall that a *Bernoulli shift* is a shift $(\Sigma(\ell), \sigma)$ together with a measure ν that is the infinite product of some probability measure on $\{1, \ldots, \ell\}$. Obviously, ν is invariant under σ; see [Orn74], [Sin76].

We will have to use the notion and properties of measure theoretic entropy, see [Sin76]. We will also need the following two theorems on subshifts of finite type.

Let A be a $k \times k$ matrix and let (Σ, σ_A) be the subshift of finite type obtained from it.

THEOREM 10.19 (Parry [Par64]). *Suppose that A^n has all its entries positive for some n. Then, there is a probability measure ν_A invariant under σ_A such that the measure theoretic entropy $h_{\nu_A}(\sigma_A)$ is equal to the topological entropy $h(\sigma_A)$. Moreover, ν_A is the only invariant probability measure having this property, and $(\Sigma_A, \sigma_A, \nu_A)$ is a mixing Markov process.*

THEOREM 10.20 (Friedman–Ornstein [Orn74]). *A mixing Markov process is isomorphic to a Bernoulli shift. In particular, the $(\Sigma_A, \sigma_A, \nu_A)$ above is Bernoulli.*

Now we begin to prove that (M, f, μ) is Bernoulli. For this, we will use the subshift (Σ_A, σ_A) and the map $\theta \colon (\Sigma_A, \sigma_A) \to (M, f)$ obtained from the Markov partition $\mathcal{R} = \{R_1, \ldots, R_k\}$.

LEMMA 10.21. *There exists $n \geq 1$ such that A^n has positive entries.*

Proof. Given R_i, we can find a periodic point $x_i \in \operatorname{int}(R_i)$; call n_i its period. Consider the unstable fiber $\mathcal{F}^u(x_i, R_i)$. For $\ell \geq 0$, we have

$$f^{\ell n_i}(\mathcal{F}^u(x_i, R_i)) \supset \mathcal{F}^u(x_i, R_i).$$

Moreover, the μ^s-length of $f^{\ell n_i}(\mathcal{F}^u(x_i, R_i))$ goes to infinity, as it is given by $\lambda^{\ell n_i} \mu^s(\mathcal{F}^u(x_i, R_i))$. This implies that

$$f^{\ell n_i}(\mathcal{F}^u(x_i, R_i)) \cap \operatorname{int}(R_j) \neq \emptyset, \qquad \forall\, j = 1, \ldots, k,$$

for ℓ large because the leaves of $\mathcal{F}^u$ are dense. Now, if

$$n = \ell \cdot \prod_{i=1}^{k} n_i$$

with ℓ large enough, we get $f^n(\operatorname{int}(R_i)) \cap \operatorname{int}(R_j) \neq \emptyset$ for each pair (i, j). Hence, we obtain that $a_{ij}^{(n)} > 0$ for each (i, j), where $A^n = (a_{ij}^{(n)})$. $\qquad\square$

This lemma shows that $(\Sigma_A, \sigma_A, \nu_A)$ is Bernoulli by the results quoted above. All we have to do now is to prove that (M, f, μ) is isomorphic to $(\Sigma_A, \sigma_A, \nu_A)$.

LEMMA 10.22. *The measure theoretic entropy $h_\mu(f)$ is $\log \lambda$.*

Proof. Since topological entropy is the supremum of measure theoretical entropies (see [Bow71, Goo71]) we have $h_\mu(f) \leq \log \lambda$. Consider now the partition $\operatorname{int}(\mathcal{R}) = \{\operatorname{int}(R_i)\}$; its μ-entropy $h_\mu(f, \operatorname{int}(\mathcal{R}))$ with respect to f is given by

$$h_\mu(f, \operatorname{int}(\mathcal{R})) = \lim_n -\frac{1}{n} \sum a_{ij}^{(n)} \lambda^{-n} y_i x_j \log(\lambda^{-n} y_i x_j)$$

where $y_i = \mu^u(\mathcal{F}^s\text{-fiber of } R_i)$ and $x_j = \mu^s(\mathcal{F}^u\text{-fiber of } R_j)$. As we saw at the end of Section 10.4,

$$\frac{a_{ij}^{(n)}}{\lambda^n} \leq \frac{\|A^n\|}{\lambda^n} \quad \text{is bounded} \ \left(\text{by } \frac{\sum y_i}{\min y_i}\right).$$

This implies

$$\lim_n -\frac{1}{n} \sum a_{ij}^{(n)} \lambda^{-n} y_i x_j \log y_i x_j = 0.$$

We also have

$$\sum a_{ij}^{(n)} \lambda^{-n} y_i x_j = \sum y_j x_j = \sum \mu(\operatorname{int}(R_j)) = \mu(M) = 1.$$

By putting these facts together, we obtain $h_\mu(f, \text{int}(\mathcal{R})) = \log \lambda$. Hence, $h_\mu(f) = \log \lambda$, since $\log \lambda = h_\mu(f, \text{int}(\mathcal{R})) \leq h_\mu(f) \leq h(f) = \log \lambda$. $\qquad \square$

Proof of Theorem 10.18. Set

$$\partial \mathcal{R} = \bigcup_{i=1}^{k} \partial R_i.$$

We have $\mu(\partial \mathcal{R}) = 0$. This implies that the set

$$Z = M - \bigcup_{i \in \mathbb{Z}} f^i(\partial R)$$

has μ-measure equal to 1. We know by Section 10.4 that θ induces a (bicontinuous) bijection of $\theta^{-1}(Z)$ onto Z. We can then define a probability measure ν on Σ_A by $\nu(B) = \mu(\theta[\theta^{-1}(Z) \cap B])$ for each Borel set $B \subset \Sigma_A$. It is easy to see that ν is σ_A invariant. Moreover, θ gives rise to a measure theoretic isomorphism between $(\Sigma_A, \sigma_A, \nu)$ and (M, f, μ). In particular $h_\nu(\sigma_A) = h_\mu(f) = \log \lambda$. Since $\log \lambda$ is also the topological entropy of σ_A, we obtain from Parry's theorem that $\nu = \nu_A$ and that $(\Sigma_A, \sigma_A, \nu)$ is a mixing Markov process. By the Friedman–Ornstein theorem, $(\Sigma_A, \sigma_A, \nu)$ is Bernoulli, hence (M, f, μ) is also Bernoulli. $\square$

Exposé Eleven

Thurston's Theory for Surfaces with Boundary

by François Laudenbach

Let M be a compact, connected surface with nonempty boundary, whose Euler characteristic is negative; for simplicity, we will limit ourselves to the case where M is orientable. Let g be the genus of M and b the number of boundary components. The Euler characteristic of M is given by

$$\chi(M) = 2 - 2g - b.$$

Thus, $\chi(M) < 0$ is equivalent to $b > 2 - 2g$. Such a surface may be cut by $3g - 3 + b$ curves into $2g - 2 + b$ pairs of pants. The excluded surfaces are S^2, T^2, D^2, and $S^1 \times [0,1]$. The pair of pants is the only surface with $\chi < 0$ and $b \leq 3 - 3g$. In what follows, we will restrict ourselves to the case $b > 3 - 3g$.

11.1 THE SPACES OF CURVES AND MEASURED FOLIATIONS

Here, S denotes the set of isotopy classes ($=$ homotopy classes) of simple curves in M that are not homotopic to a point or to a boundary component. Also, we consider the set $\mathcal{MF}$ of Whitehead classes of measured foliations, which are subject to the condition that each boundary curve is a cycle of leaves containing at least one singularity. Recall that Whitehead equivalence is generated by the following operations and their inverses:

- isotopy, free on the boundary;

- contraction (to a point) of a leaf in the interior joining two singularities, at most one of which is on the boundary;

- contraction (to a point) of a leaf in the boundary joining two singularities.

As in the case of surfaces without boundary, the geometric intersection number gives rise to maps

$$i_* : S \to \mathbb{R}_+^S$$

and

$$I_* : \mathcal{MF} \to \mathbb{R}_+^S.$$

Let π be the projection onto the projective space

$$\pi \colon \mathbb{R}_+^{\mathcal{S}} - \{0\} \to P(\mathbb{R}_+^{\mathcal{S}}).$$

We denote by $\mathcal{PMF}$ the image of $\pi \circ I_*$.

THEOREM 11.1. *For a compact, connected surface M with $\chi(M) < 0$, we have:*
1. *The maps i_* and I_* are injective.*
2. *$\mathcal{PMF}$ is homeomorphic to the sphere $S^{6g-7+2b}$.*
3. *The image $\pi \circ i_*(\mathcal{S})$ is dense in $\mathcal{PMF}$.*

Proof. The proof is very close to the proof in the case without boundary (see Exposés 4 and 6); we only give an explanation for the dimension.

Consider in M a system of $3g - 3 + b$ simple curves K_i that partition M into pairs of pants such that each K_i belongs to two distinct pairs of pants; such a system does not exist if M is a one-holed torus ($3g - 3 + b = 1$); we will revisit this case at the end. Once put into normal form with respect to this decomposition, a foliation $(\mathcal{F}, \mu)$ is characterized (up to equivalence) by triples (m_i, s_i, t_i), $i = 1, \ldots, 3g - 3 + b$, belonging to the boundary $\partial(\nabla \leq)$ of the triangle inequality. We have

$$m_i = I(\mathcal{F}, \mu; [K_i]).$$

Because the curves of the boundary have measure zero, the m_i determine the foliations in each pair of pants, up to equivalence. Thus, by the theory of the *pants seam*, the pair (s_i, t_i) describes how to glue the foliations in the two pairs of pants adjacent to K_i.

Finally, the set of equivalence classes of measured foliations in normal form with respect to the given decomposition of M is in bijection with a punctured positive cone, on the base $S^{6g-7+2b}$.

To obtain the theorem, it remains to show that s_i and t_i are determined by $I_*(\mathcal{F}, \mu)$ and that the image $I_*(\mathcal{MF})$ is a topological manifold. These two points are proven as in the case without boundary. To be precise, s_i and t_i are calculated with the aid of the measures of classes $[K_i']$ and $[K_i'']$ associated with the decomposition (see Exposé 6); it suffices then to remark that these classes are truly elements of $\mathcal{S}$.

In the case $3g - 3 + b = 1$ (M is a one-holed torus), we take for K_1 and K_1' two "generators" of the torus and for K_1'' the curve obtained from K_1' by a positive Dehn twist along K_1 (see Figure 11.1). We leave to the reader the exercise of establishing the formulas that give s_1 and t_1 as functions of the measures of $[K_1]$, $[K_1']$, and $[K_1'']$.

$\square$

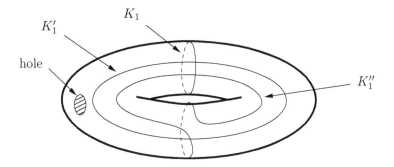

Figure 11.1. The curves K_1, K_1', K_1'' on the punctured torus M

11.2 TEICHMÜLLER SPACE AND ITS COMPACTIFICATION

We consider the topological space $\mathcal{H}$ of Riemannian metrics of curvature -1, for which each boundary curve is a geodesic of length 1. The group $\text{Diff}_0(M)$, the group of diffeomorphisms of M isotopic to the identity, acts naturally on $\mathcal{H}$. We define the *Teichmüller space*[1] of M to be the topological quotient space

$$\mathcal{T} = \mathcal{H}/\text{Diff}_0(M).$$

We parametrize $\mathcal{T}$ by fixing a pair of pants decomposition as in the proof of Theorem 11.1. A Teichmüller structure (i.e., a point of $\mathcal{T}$) is completely determined by the lengths m_i of the geodesics isotopic to the curves K_i and by the "angles" (real numbers) θ_i given by the gluing. One shows, via this parametrization, that $\mathcal{T}$ is homeomorphic to $(\mathbb{R}_+^* \times \mathbb{R})^{3g-3+b}$.

Further, for each $\alpha \in \mathcal{S}$, we may speak of the length with respect to the Teichmüller structure in consideration. We thus have a map

$$\ell_* \colon \mathcal{T} \to \mathbb{R}_+^{\mathcal{S}}$$

As in the case without boundary, the "angle" α_i is determined by the lengths of the geodesics of $[K_i']$ and $[K_i'']$. We therefore have the following theorem.

THEOREM 11.2. *The Teichmüller space being defined as above, the map ℓ_* is a proper function which is a homeomorphism onto its image. In particular, $\ell_*(\mathcal{T})$ is homeomorphic to $\mathbb{R}^{6g-6+2b}$.*

From here on, we will identify $\mathcal{MF}$ and $\mathcal{T}$ with their respective images in $\mathbb{R}_+^{\mathcal{S}}$.

LEMMA 11.3. *In $\mathbb{R}_+^{\mathcal{S}}$, the spaces $\mathcal{MF}$ and $\mathcal{T}$ are disjoint.*

---

[1]Classically [Har77], one does not fix the length of the curves of the boundary.

Proof. It suffices, for example, to find for each foliation $(\mathcal{F}, \mu)$ a sequence $\alpha_n \in \mathcal{S}$ such that $I((\mathcal{F}, \mu); \alpha_n) \to 0$. Let $q : \widetilde{M} \to M$ be the ramified covering of transverse orientations of $\mathcal{F}$. Let $\widetilde{\mathcal{F}} = q^*(\mathcal{F})$. Let $z \in \text{int } \widetilde{M} \setminus \text{Sing } \widetilde{\mathcal{F}}$ be a limit point for a leaf L (a point of recurrence). We may form a simple curve C_n from an arc of L and a transverse arc of measure at most $1/n$ (we choose this arc in a "flow box" neighborhood of z). As $\widetilde{\mathcal{F}}$ is transversely orientable, C_n can be approximated by a true transversal to $\widetilde{\mathcal{F}}$; we may suppose in addition that any double points of $q(C_n)$ are isolated. By a modification around each double point, we construct a curve C'_n that is a simple curve in M, transverse to $\mathcal{F}$ and of measure $\leq \frac{1}{n}$. We set $\alpha_n = [C'_n]$.

It remains to show that C'_n is not isotopic to a curve of the boundary. If it were, we would have an annulus equipped with a measured foliation, where one boundary curve is transverse to the foliation and the other is a cycle; this is forbidden by Poincaré recurrence (Theorem 5.2).

$\square$

THEOREM 11.4. *The projection π injects $\mathcal{T}$ into $P(\mathbb{R}_+^{\mathcal{S}})$, and $\pi|\mathcal{T}$ is a homeomorphism of $\mathcal{T}$ onto its image, which is disjoint from $\mathcal{PMF}$. Endowed with the induced topology, $\pi(\mathcal{T}) \cup \mathcal{PMF}$ is a manifold with boundary $\overline{\mathcal{T}}$, and it is homeomorphic to a ball of dimension $6g - 6 + 2b$. The mapping class group $\pi_0(\text{Diff}(M))$ acts continuously on $\overline{\mathcal{T}}$.*

For the proof, we follow the same procedure as in the case without boundary (Exposé 8) and not the one suggested by the order of the sentences in the statement of the theorem.

11.3   A SKETCH OF THE CLASSIFICATION OF DIFFEOMORPHISMS

We emphasize here that we are dealing with a classification of diffeomorphisms up to isotopy, where the isotopy on the boundary is free.

Let $\varphi \in \text{Diff}(M)$ and let $[\varphi]$ be its isotopy class. By the Brouwer fixed point theorem, there exists a point $x \in \overline{\mathcal{T}}$ such that

$$[\varphi] \cdot x = x.$$

If $x \in \mathcal{T}$, then φ is isotopic to a hyperbolic isometry of x; in this case, $[\varphi]$ is of finite order (cf. Exposé 9).

If $x \in \mathcal{PMF}$, there is a foliation $(\mathcal{F}, \mu)$ and a $\lambda > 0$ such that

$$\varphi(\mathcal{F}, \mu) \sim (\mathcal{F}, \lambda\mu),$$

where the equivalence is in the sense of Whitehead. From this point on, everything depends on $(\mathcal{F}, \mu)$ and λ.

Let Σ be the complex consisting of the singularities and the leaves joining two singularities (possibly joining some singularity to itself). The complex Σ contains ∂M since each component of the boundary contains a singularity. We denote by $U(\mathcal{F})$ the complement of a regular neighborhood of Σ in M. We see that, up to isotopy, $U(\mathcal{F})$ only depends on the Whitehead class of $\mathcal{F}$.

We define $\beta U(\mathcal{F})$ as the union of the boundary components of $U(\mathcal{F})$ that represent elements of $\mathcal{S}$. We distinguish the following cases:

1. $\beta U(\mathcal{F}) \neq \emptyset$ (reducible case).

2. $\beta U(\mathcal{F}) = \emptyset$ and $\lambda = 1$ (periodic case).

3. $\beta U(\mathcal{F}) = \emptyset$ and $\lambda \neq 1$ (pseudo-Anosov case).

Reducible diffeomorphism. We say that φ is *reducible* if there exist mutually disjoint simple closed curves $\gamma_1, \ldots, \gamma_n$, representing distinct elements of $\mathcal{S}$, such that $\varphi(\gamma_1 \cup \cdots \cup \gamma_n) = \gamma_1 \cup \cdots \cup \gamma_n$.

LEMMA 11.5. *If $\varphi(\mathcal{F}, \mu) \sim (\mathcal{F}, \lambda\mu)$ and if $\beta U(\mathcal{F})$ is not empty, then φ is isotopic to a reducible diffeomorphism.*

Proof. Up to changing φ by an isotopy, we may suppose that $\varphi(U(\mathcal{F})) = U(\mathcal{F})$. Let γ_1 be a component of $\beta U(\mathcal{F})$, and let $\gamma_{i+1} = \varphi(\gamma_i)$ for $i = 1, 2, \ldots$ We stop at $\gamma_n = \varphi^{n-1}(\gamma_1)$ if it is the first iterate such that $\varphi(\gamma_n)$ is isotopic to γ_1. Since $\varphi(\gamma_1)$ and γ_1 bound an annulus, it is not difficult to produce an isotopy of φ for which $\gamma_1 \cup \cdots \cup \gamma_n$ is invariant. $\square$

If we cut M along $\gamma_1, \ldots, \gamma_n$, we obtain a "simpler" surface $\widehat{M}$ on which φ induces a diffeomorphism. Observe that each component of $\widehat{M}$ is either a pair of pants or satisfies $b > 3g - 3$; indeed, no two curves among the γ_i are isotopic, and no γ_i is isotopic to a curve of the boundary. Note that the number of possible successive reductions has an upper bound depending only on M; when all of the pieces are pairs of pants, no further reduction is possible. (A small difficulty arises because $\widehat{M}$ is not, in general, connected; we will revisit this in Section 11.4.)

Arational foliations. If $\beta U(\mathcal{F})$ is empty, we say that $\mathcal{F}$ is *arational*. There is then a distinguished representative in the class of $\mathcal{F}$, where there are no connections between singularities of the interior (neither among themselves nor with those of the singularities on the boundary) and where the singularities on the boundary are simple (a single separatrix enters into the interior). This representative is unique up to isotopy. In what follows, we suppose that $\mathcal{F}$ is this canonical representative. The equivalence $\varphi(\mathcal{F}, \mu) \sim (\mathcal{F}, \lambda\mu)$ therefore gives rise to an equality:

$$\varphi(\mathcal{F}, \mu) = (\mathcal{F}, \lambda\mu),$$

under the condition that we may have to modify φ by a suitable isotopy.

To each system τ of arcs transverse to $\mathcal{F}$ there is an associated system of $\mathcal{F}$-rectangles. The union of these rectangles is a subset N of M, whose frontier is a union of cycles of leaves. Since $\beta U(\mathcal{F})$ is empty, the frontier of N is in ∂M, thus $N = M$. From this, we deduce that each half-leaf that does not meet a singularity is everywhere dense.

Case 1. $\varphi(\mathcal{F}, \mu) = (\mathcal{F}, \mu)$

As in the case of closed surfaces, we consider a "good" system of transverse arcs τ (see Exposé 9). Up to changing φ by an isotopy that preserves $\mathcal{F}$, we reduce to the case where $\varphi(\tau) = \tau$ and thus where φ preserves the system of rectangles; from this we deduce that φ is isotopic to a periodic diffeomorphism.

Case 2. $\varphi(\mathcal{F}, \mu) = (\mathcal{F}, \lambda\mu)$, $\lambda > 1$

With the aim of constructing a second invariant foliation, we have to modify the construction of a "good" system of transverse arcs given in Lemma 9.9. In each sector of an interior singularity of M, we take a small arc transverse to $\mathcal{F}$; however, we do not put any in the sectors adjacent to the boundary. Also, for each smooth leaf of the boundary, we choose a point that we make an endpoint of a small transversal that enters the interior (Figure 11.2).

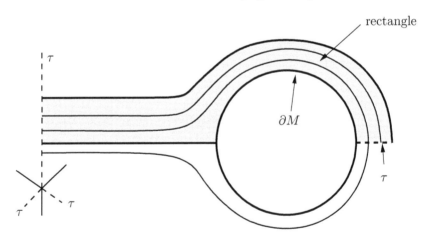

Figure 11.2.

If τ is such a system of arcs, then, by a suitable isotopy of φ along the leaves of $\mathcal{F}$, we obtain $\varphi(\tau) \subset \tau$. From this, the techniques of Lemma 9.9 and Lemma 9.11 apply to construct a pre-Markov partition $\{R_i\}$.

Let a_{ij} be the number of components of $\varphi(\text{int } R_i) \cap \text{int } R_j$. Let x_i be the μ-measure of $\partial_\tau^0 R_i$ (or of $\partial_\tau^1 R_i$). We have

$$\lambda x_j = \sum_i x_i a_{ij}.$$

In other words, if A is the matrix (a_{ij}), the column vector (x_i) is an eigenvector of the transpose matrix A^t, with eigenvalue λ. By the same proof as in the case without boundary, we prove that A also has an eigenvector (y_i), whose coordinates are all strictly positive, with an eigenvalue of $1/\xi > 0$:

$$y_i = \xi \sum a_{ij} y_j.$$

Observe that the geometric proof (in the case without boundary) rests on the fact that, for each i, $\cup \varphi^n(\partial_{\mathcal{F}}^0 R_i)$ is dense. This is again true here because $\partial_{\mathcal{F}}^0 R_i$ cannot be entirely contained in the boundary of M; it necessarily contains an arc of a leaf of the interior.

Construction of the foliation $(\mathcal{F}', \mu')$. As in the case without boundary, we start by fixing the μ'-measure of the arcs $\partial_{\mathcal{F}}^\epsilon R_i$, $\epsilon \in \{0, 1\}$. If such an arc is in ∂M, we assign it measure 0, because we want ∂M to also be a union of cycles of leaves for $\mathcal{F}'$ (Figure 11.3).

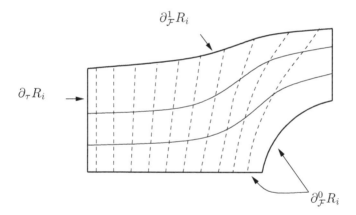

Figure 11.3.

Now, we draw in each rectangle R_i a measured foliation $(\mathcal{F}', \mu')$ that is transverse to $\mathcal{F}$ and that respects the assigned measures; this condition guarantees that we can glue the pieces together. We observe that

$$\mathrm{Sing}\, \mathcal{F}' \cap \mathrm{int}\, M = \mathrm{Sing}\, \mathcal{F} \cap \mathrm{int}\, M,$$

while the singularities of $\mathcal{F}$ on ∂M become regular points of $\mathcal{F}'$; we have

$$\mathrm{Sing}\, \mathcal{F}' \cap \partial M = \tau \cap \partial M.$$

Now that we have a measure μ' on the leaves of $\mathcal{F}$, we construct a pseudo-Anosov diffeomorphism φ' that respects the two foliations, dilating the leaves of $\mathcal{F}$ by $1/\xi$ and contracting those of $\mathcal{F}'$ by $1/\lambda$. We have thus proved that $\xi = 1/\lambda$. Note that φ' is the identity on the boundary.

Pseudo-Anosov diffeomorphism.   We say that φ is a *pseudo-Anosov diffeo-morphism* if there exist two invariant measured foliations, $(\mathcal{F}^s, \mu^s)$ and $(\mathcal{F}^u, \mu^u)$, and a $\lambda > 1$ with the following properties:

1. $\varphi(\mathcal{F}^s, \mu^s) = (\mathcal{F}^s, \frac{1}{\lambda}\mu^s)$.

2. $\varphi(\mathcal{F}^u, \mu^u) = (\mathcal{F}^u, \lambda\mu^u)$.

3. $\mathcal{F}^s$ and $\mathcal{F}^u$ are transverse at each point of the interior.

4. Each component of ∂M is a cycle of leaves of $\mathcal{F}^s$ and of $\mathcal{F}^u$ and contains singularities of these two foliations, and φ is the identity on the boundary.

N.B. φ is not C^1 along the boundary.

The properties of pseudo-Anosov diffeomorphisms indicated in Section 9.6 are still true. Only Proposition 9.21 requires a modification: it applies only to isotopy classes of curves not homotopic to a component of the boundary; for that matter, the metric $\sqrt{(d\mu^s)^2 + (d\mu^u)^2}$ is singular along the whole boundary.

Example: the disk with three holes.   Let A be an Anosov matrix acting on T^2. Let σ be the involution $(x, y) \mapsto (-x, -y)$; it has four fixed points. We may regard $T^2 \to T^2/\sigma$ as a ramified covering. As can be seen by calculating the Euler characteristic, the base is a 2-sphere.

The transformation A leaves invariant two linear foliations of irrational slope which therefore pass to the quotient, inducing on S^2 two measured foliations $(\mathcal{F}^s, \mu^s)$ and $(\mathcal{F}^u, \mu^u)$, with singularities at the four ramification points (on T^2, the transverse measures are given by closed 1-forms with constant coefficients that define the respective foliations). Since the degree of ramification is 2, the singularities are of the type indicated in Figure 11.4.

Since A commutes with σ, A induces on S^2 a homeomorphism φ that leaves invariant $\mathcal{F}^u$ and $\mathcal{F}^s$ and that transforms the measures in the same way as A on T^2. To obtain the disk with three holes, equipped with a pseudo-Anosov diffeomorphism, we blow up the singularities as shown in Figure 11.5.

11.4   THURSTON'S CLASSIFICATION AND NIELSEN'S THEOREM

The arguments of the previous subsection lead, at least in the orientable case, to a proof of the following theorem.

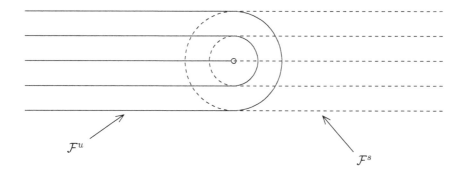

Figure 11.4.

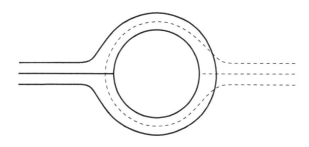

Figure 11.5. Blowing up singularities

THEOREM 11.6. *For any diffeomorphism φ of a compact, connected surface satisfying $b > 2 - 2g$, φ is isotopic to φ' having one of the following three properties:*

(1) φ' is of finite order, hence an isometry of a hyperbolic structure.

(2) φ' is pseudo-Anosov.

(3) φ' is reducible.

To pursue the analysis in the reducible case, it is necessary to figure out how to work with a disconnected surface. We only need to consider the case where φ acts transitively on $\pi_0(M)$. Therefore, let $M = M_1 \cup \cdots \cup M_n$ where the M_i are the connected components of M, and say $\varphi(M_i) = M_{i+1}$ for $i = 1, \ldots, n-1$ and $\varphi(M_n) = M_1$.

We will say that φ is pseudo-Anosov if $\varphi^n|_{M_1}$ is pseudo-Anosov; then, by conjugation, we see that $\varphi^n|_{M_i}$ is pseudo-Anosov for all i. If one knows that $\varphi^n|_{M_1}$ is isotopic to a pseudo-Anosov diffeomorphism (resp. a diffeomorphism of finite order), then φ is isotopic to such a diffeomorphism (on each M_i, take the foliation associated to φ^n); it suffices to read the isotopy of $\varphi^n|_{M_1}$ as an isotopy of $\varphi : M_n \to M_1$. In this way, one obtains the final result below, in which one can avoid restrictions on the genus or the Euler characteristic, since the cases of S^2, T^2, D^2, the pair of pants, the Möbius band, and the Klein bottle are known.

THEOREM 11.7. *Let φ be a diffeomorphism of a compact surface M. There exist (possibly disconnected) compact surfaces $M_1, \ldots, M_k$ with the following properties:*

1. $M = M_1 \cup \cdots \cup M_k$, that is, there are inclusions $M_i \to M$ that cover M.

2. The inclusions are injective on the interiors of the M_i, but two boundary components of the M_i might map to the same closed curve; we denote by $C_1, \ldots, C_r$ these "cutting" curves.

3. For $i \neq j$, C_i is not isotopic to C_j.

4. φ is isotopic to a diffeomorphism φ' that preserves the images of the M_i for all i, and restricts to the identity map on the C_i.

5. The induced map of φ' on M_i is isotopic in $\mathrm{Diff}(M_i)$ to a periodic diffeomorphism or to a pseudo-Anosov diffeomorphism.

So, paradoxically, if φ is a Dehn twist along a curve C, this classification drops mention of φ entirely. That is, if one cuts M along C and allows a free isotopy on the boundary, one arrives at the identity.

THEOREM 11.8 (Nielsen [Nie43]). *Let φ be a diffeomorphism of a compact surface representing an element of order n in the mapping class group $\pi_0(\mathrm{Diff}(M))$. Then φ is isotopic to a periodic diffeomorphism of order n.*

Proof. We limit ourselves to the nontrivial case $b > 3 - 3g$. Let $M_1 \cup \cdots \cup M_k$ be a decomposition of the surface as in the preceding theorem. For each cutting curve C_i, consider a small tubular neighborhood $N_i = C_i \times [-1, 1]$. We denote by M'_j the part of M_j that remains after removing the open collars. After the first isotopy, we have $\varphi(M'_j) = M'_j$ for each j and $\varphi(C_i \times \{t\})$ is of the form $C'_i \times \{t'\}$. Furthermore, $\varphi|_{M'_j}$ is periodic or pseudo-Anosov.

Claim 1: The diffeomorphism φ^n preserves each C_i with its orientation and normal orientation.

Proof of Claim 1: If $i \neq j$, C_i is not isotopic to C_j. Also, C_i cannot be isotoped to its opposite except on the Klein bottle (excluded by hypothesis). Finally, an exchange of sides induces a nontrivial morphism on $H_1(M, \mathbb{Z})$.

Claim 2: The isotopy of φ^n to the identity can be chosen through diffeomorphisms that preserve $C_1 \cup \cdots \cup C_r$.

Proof of Claim 2: An isotopy from φ^n to the identity induces a loop, based at C_1, in the space of simple curves on M. Since M is not the torus or the Klein bottle, such a loop is homotopic to a point. (See [Gra73]).

By lifting this homotopy to $\mathrm{Diff}(M)$, we find an isotopy of φ^n to the identity in $\mathrm{Diff}(M, C_1)$. We proceed in the same fashion for the other curves.

Consequently, for each j, $\varphi^n|_{M'_j}$ is isotopic to the identity in $\mathrm{Diff}(M'_j)$. Thus $\varphi|_{M'_j}$ cannot be pseudo-Anosov; hence $\varphi|_{M'_j}$ is periodic and, since it is an isometry for a particular hyperbolic metric[2] (see Exposé 9), $\varphi^n|_{M'_j}$ is the identity. Thus $\varphi^n|_{N_i}$ is a certain iterate θ^{q_i} of a Dehn twist θ along the curve C_i.

Claim 3: The integer q_i is zero for each i.

Proof of Claim 3: There exists a class $\beta \in \mathcal{S}$ such that $i(\beta, [C_i]) \neq 0$ and $i(\beta, [C_j]) = 0$ for $j \neq i$; if q_i is not zero, then by Appendix A we have $i(\varphi^n(\beta), \beta) \neq 0$, which is forbidden since φ^n is isotopic to the identity.

Suppose that $\varphi(C_i) = C_i$. Claim 3 then implies that φ makes the two boundaries of the collar N_i turn in the same direction. More precisely, in suitable coordinates, $\varphi(x, \pm 1) = (x + \frac{1}{n}, \pm 1)$, where $x \in \mathbb{R}/\mathbb{Z}$, and $\varphi(\{0\} \times [-1, 1])$ is isotopic to $\{\frac{1}{n}\} \times [-1, +1]$, rel boundary. From here, it is easy to make an isotopy of $\varphi|_{N_i}$, trivial along ∂N_i, to a periodic diffeomorphism of period n. If $\varphi(C_i) \neq C_i$, we proceed in the same fashion with the orbit of C_i. $\square$

Remark. Nielsen's proof of Theorem 11.6 rests on the fact that φ lifts in the universal cover to a $\widetilde{\varphi}$ that extends to the boundary of the Poincaré disk; $\widetilde{\varphi}|_{\partial \mathbb{D}^2}$ depends only on the homotopy class of φ. Underlying the proof that we have given here is a different compactification, namely, that of Teichmüller space. On the other hand, as Fenchel announced (see [Fen50] or the book of Fenchel and Nielsen [FN03]), we may deduce from the Smith fixed point theorem [Smi34] that, if G is a finite solvable subgroup of the mapping class group $\pi_0(\mathrm{Diff}(M))$, then G admits a fixed point in (uncompactified) Teichmüller space; from this we may deduce that G lifts to a subgroup of $\mathrm{Diff}(M)$.

The argument, briefly, is as follows. Let $F \to G \to \mathbb{Z}/p\mathbb{Z}$ be an extension where p is a prime number, and suppose that the result holds for F. Let $\mathcal{T}_F$ be the set of fixed points of F in $\mathcal{T}$. Let $M' = M/F$, for a chosen action of F on M, and let X be the set of ramification points. Let $\mathcal{T}(M', X)$ be the set of conformal structures of M' modulo the identity component of $\mathrm{Diff}(M', X)$. We show that this is a cell (using the theorem of Earle and Eells [EE69] that the action of $\mathrm{Diff}_0(M')$ on the metrics of curvature -1 gives the structure of a principal fibration and the fact that $\mathrm{Diff}_0(M')/\mathrm{Diff}_0(M', X)$ is contractible; for example, if X is a single point, this last quotient is homeomorphic to the universal cover $\widetilde{M}$).

---

[2] The hyperbolic metric obtained by this argument may not have boundary curves of length 1; hence we do not say here that $\varphi|_{M'_j}$ admits a fixed point in $\mathcal{T}(M'_j)$. Question: is there a Teichmüller metric invariant for $\varphi|_{M'_j}$?

Furthermore, we show that $\mathcal{T}_F$ is homeomorphic to $\mathcal{T}(M', X)$; thus $\mathcal{T}_F$ is also a cell. Finally, as F is invariant in G, it follows that G acts on $\mathcal{T}_F$ via the quotient $\mathbb{Z}/p\mathbb{Z}$. By the Smith fixed point theorem, there is a fixed point.[3]

11.5   THE SPECTRAL THEOREM

For a Riemannian metric ρ and a simple curve c, denote the length of c by $L_\rho(c)$, and let $[c]$ denote its isotopy class. We set

$$\ell_\rho([c]) = \inf\{L_\rho(c') \mid c' \text{ is isotopic to } c\}.$$

THEOREM 11.9. *For each diffeomorphism φ of a compact surface M, there exists a finite sequence $\lambda_1, \ldots, \lambda_k \geq 1$ such that, for each $\alpha \in \mathcal{S}$ and for any Riemannian metric ρ, the sequence $\sqrt[n]{\ell_\rho(\varphi^n(\alpha))}$ converges to a limit that is independent of ρ and that belongs to $\{\lambda_1, \ldots, \lambda_k\}$. The numbers $\lambda_1, \ldots, \lambda_k$ are algebraic integers whose degrees admit an upper bound only depending on the Euler characteristic of M.*

Proof. By Theorem 11.7, we may suppose that $M = M_1 \cup \cdots \cup M_k$, that $\varphi(M_i) = M_i$, and that $\varphi|_{M_i}$ is isotopic in $\mathrm{Diff}(M_i)$ to a diffeomorphism φ_i, where φ_i is pseudo-Anosov with dilatation factor λ_i for $i = 1, \ldots, m$, and where φ_i is periodic for $i = m + 1, \ldots, k$. For $i > m$, we set $\lambda_i = 1$. We will prove that this "spectrum" satisfies the statement of the theorem.

Since all Riemannian metrics are equivalent, we may restrict ourselves to the case where ρ is a hyperbolic metric that admits the ∂M_i as geodesics. Then, the geodesic c in the class α intersects ∂M_i minimally. We cut c into arcs $c_1, \ldots, c_r$ corresponding to the different segments of c in the M_i: $c_s \subset M_{i(s)}$. By Section 3.3, c_s is an essential arc in $M_{i(s)}$ (nontrivial in $\pi_1(M_{i(s)}, \partial M_{i(s)})$); hence, $\varphi(c_s)$ is also an essential arc in $M_{i(s)}$. Furthermore, the geodesic of the class $\varphi^n(\alpha)$ is $c_1^{(n)} \cup \cdots \cup c_r^{(n)}$, where $c_s^{(n)}$ is isotopic to $\varphi^n(c_s)$ by an ambient isotopy in $\mathrm{Diff}(M_{i(s)})$.

Let us say that $\lambda_{i(1)} \geq \lambda_{i(s)}$ for $s = 1, \ldots, r$. We will prove that

$$\sqrt[n]{\ell_\rho(\varphi^n(\alpha))} \to \lambda_{i(1)}.$$

For fixed α, all of the classes $\varphi^n(\alpha)$ traverse the same M_i. Thus, it suffices to prove the statement for a subsequence $(\varphi^t)^n$. This allows us to reduce to the case where φ is the identity on the ∂M_i and where $\lambda_{i(s)} = 1$ implies $\varphi_{i(s)}$ is equal to the identity, as we shall assume in what follows.

Consider the geodesic arc $d_s^{(n)}$ (resp. $h_s^{(n)}$) that is homotopic to $\varphi^n(c_s)$ with endpoints fixed (resp. free). Let $\beta_s^{(n)}$ (resp. $\delta_s^{(n)}$) be the shortest path joining

---

[3]I wish to thank Alexis Marin, who communicated to me the essential elements of this remark.

the origin (resp. the endpoint) of $d_s^{(n)}$ to that of $h_s^{(n)}$ so that $d_s^{(n)}$ is homotopic, with endpoints fixed, to $\beta_s^{(n)} * h_s^{(n)} * \left[\delta_s^{(n)}\right]^{-1}$. Let us provisionally accept the following.

Claim. The growth of $L_\rho\left(\beta_s^{(n)}\right)$ and $L_\rho\left(\delta_s^{(n)}\right)$ is subexponential; that is to say

$$\limsup \frac{1}{n} \log \left(L_\rho\left(\beta_s^{(n)}\right)\right) = \limsup \frac{1}{n} \log \left(L_\rho\left(\delta_s^{(n)}\right)\right) = 0.$$

If $\lambda_{i(s)} = 1$, it is clear that $L_\rho\left(h_s^{(n)}\right)$ is bounded. If $\lambda_{i(s)} > 1$, then

$$\sqrt[n]{L_\rho\left(h_s^{(n)}\right)} \to \lambda_{i(s)};$$

indeed $\varphi_{i(s)}$ is pseudo-Anosov with dilatation coefficient $\lambda_{i(s)}$ and, in this case, the result is given by Proposition 9.21 (with one slight difference that here the manifold has boundary and that it acts on the free isotopy classes of paths going from boundary to boundary, but the proof is the same). By the claim, we have

$$\sqrt[n]{L_\rho\left(d_s^{(n)}\right)} \to \lambda_{i(s)}.$$

In addition, we have the following inequalities:

$$\sum_s L_\rho\left(h_s^{(n)}\right) \leq \sum_s L_\rho\left(c_s^{(n)}\right) = \ell_\rho\left(\varphi^n(\alpha)\right) \leq \sum_s L_\rho\left(d_s^{(n)}\right).$$

Considering the growth of each term, we find that $\sqrt[n]{\ell_\rho(\varphi^n(\alpha))}$ tends to $\lambda_{i(1)}$.

Proof of claim. We know, by a theorem of Lickorish (see [Lic64]), that $\varphi|_{M_{i(s)}}$ is isotopic rel boundary to a composition of Dehn twists along simple curves in $M_{i(s)}$ (see Exposé 15). We consider therefore the situation of a surface with boundary N, endowed with a hyperbolic metric ρ, and a twist ψ along a geodesic α in N. Let c be the minimizing geodesic for a nontrivial class of $\pi_1(N, \partial N)$; let h be the minimizing geodesic in the class of $\psi(c)$. In the homotopy of $\psi(c)$ to h, each endpoint of the arc shifts along the boundary: we have on ∂N geodesic arcs β and δ, such that $\psi(c)$ is homotopic to $\beta * h * \delta^{-1}$, with endpoints fixed. Say that each component of ∂N has length equal to 1; then the claim, and hence the theorem is a consequence of the following statement:

$$L_\rho(\beta), L_\rho(\delta) \leq 1.$$

So we are reduced to proving this inequality. Note that if α is isotopic to a component of the boundary, there is nothing to show since $h = c$ and the displacement of each endpoint of $\psi(c)$ in the course of its isotopy to c is exactly one turn.

In general, the intersection of h with c is minimal in the free homotopy class of h. Let c' be an arc parallel to c. If $\operatorname{card}(\psi(c') \cap c) = \operatorname{card}(h \cap c)$, then $\psi(c')$ and h are isotopic by an isotopy that leaves c invariant (Proposition 3.13). In this case, the displacement of the endpoints during this isotopy is less than one turn.

By Appendix A, we see that $\operatorname{card}(\psi(c') \cap c)$ does not decrease as long as the endpoints of $\psi(c')$ remain fixed. On the other hand, by shifting the origin of $\psi(c')$ on top of that of c, we possibly reduce $\operatorname{card}(\psi(c') \cap c)$; we say that we "drive" a point of intersection to the boundary. To shift the origin of $\psi(c')$ by more than one turn requires an immersion of a triangle in N, as indicated in Figure 11.6, in the domain of the immersion. From this, we deduce that α is isotopic to a component of the boundary, which we have excluded at the beginning. Since the shift of the origin of c to that of c' is arbitrarily small, we finally have that the shift of the origin of $\psi(c)$ to h is less than one turn. This completes the proof of the inequality, hence the claim, and hence the theorem. $\square$

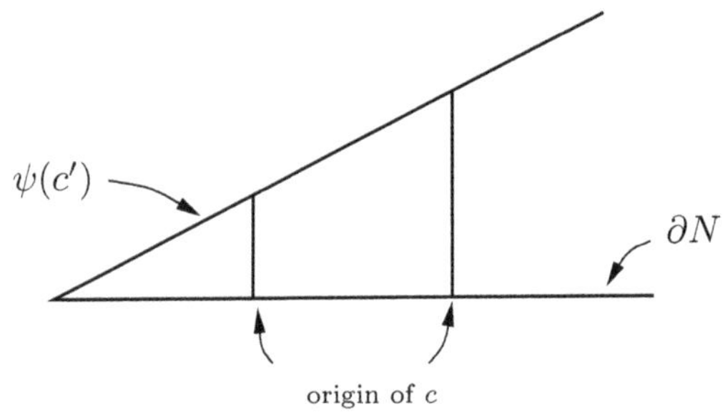

Figure 11.6.

Exposé Twelve

Uniqueness Theorems for Pseudo-Anosov Diffeomorphisms

by Albert Fathi and Valentin Poénaru

12.1 STATEMENT OF RESULTS

In what follows, M is a closed, orientable surface of genus $g \geq 2$. We are given a pseudo-Anosov diffeomorphism $\varphi : M \to M$. This means that there are two transverse measured foliations $(\mathcal{F}^s, \mu^s)$ and $(\mathcal{F}^u, \mu^u)$ and a number $\lambda > 1$ such that

$$\varphi(\mathcal{F}^s, \mu^s) = \left(\mathcal{F}^s, \frac{1}{\lambda}\mu^s \right)$$

and

$$\varphi(\mathcal{F}^u, \mu^u) = (\mathcal{F}^u, \lambda\mu^u)$$

(i.e., φ contracts distances between the leaves of $\mathcal{F}^u$ by a factor of $1/\lambda$).

We recall what unique ergodicity means. First of all, there is an $\mathcal{F}^s$-invariant measure $\mu(= \mu^s)$ defined on each transversal T to (the nonsingular part of) $\mathcal{F}^s$; this is a Borel measure μ_T that is finite on each compact transversal and that is invariant under (the germs of) the holonomy of $\mathcal{F}^s$. The foliation $\mathcal{F}^s$ is *uniquely ergodic* if there exists a single $\mathcal{F}^s$-invariant measure up to multiplication by a scalar, that is:

- there exists a measure μ invariant under $\mathcal{F}^s$ and

- if ν is another measure invariant under $\mathcal{F}^s$, there exists a scalar $\lambda \in \mathbb{R}$ such that $\nu_T = \lambda\mu_T$ for every transversal T.

THEOREM 12.1 (Unique ergodicity). *The stable and unstable foliations of a pseudo-Anosov diffeomorphism are uniquely ergodic.*

Theorem 12.1 is a particular case of a result of Bowen and Marcus [BM77].

We recall that a pseudo-Anosov diffeomorphism gives a natural invariant positive measure, determined up to a positive constant. This measure is given locally by the product of μ^s and μ^u. Up to multiplying μ^s (or μ^u) by a constant, we can suppose that $\mu^s \otimes \mu^u$ is a probability measure, that is, $\mu^s \otimes \mu^u(M) = 1$.

THEOREM 12.2. *Let φ be a pseudo-Anosov diffeomorphism, and suppose that $\mu^s \otimes \mu^u(M) = 1$. If $\alpha, \beta \in \mathcal{S}$, we have*

$$\lim_{n\to\infty} \frac{i(\varphi^n(\alpha), \beta)}{\lambda^n} = I(\mathcal{F}^s, \mu^s; \alpha) I(\mathcal{F}^u, \mu^u; \beta).$$

COROLLARY 12.3. *If $\alpha \in \mathcal{S}$, and if $[\alpha]$, $[\mathcal{F}^s, \mu^s]$, and $[\mathcal{F}^u, \mu^u]$ are the images of α, $(\mathcal{F}^s, \mu^s)$, and $(\mathcal{F}^u, \mu^u)$ in $\mathcal{PMF}$, we have*

$$\lim_{n\to\infty} [\varphi^n(\alpha)] = [\mathcal{F}^u, \mu^u] \quad and$$
$$\lim_{n\to\infty} [\varphi^{-n}(\alpha)] = [\mathcal{F}^s, \mu^s].$$

In fact, Thurston gives the following stronger result: if $[\mathcal{F}, \mu] \in \mathcal{PMF}$ and if $[\mathcal{F}, \mu] \neq [\mathcal{F}^s, \mu^s]$, then

$$\lim_{n\to\infty} \varphi^n([\mathcal{F}, \mu]) = [\mathcal{F}^u, \mu^u].$$

It is possible that our proof of Theorem 12.2 can also recover this stronger result, by doing uniform estimates of convergence on compact sets of $\mathcal{PMF} - \{[\mathcal{F}^s, \mu^s]\}$.

COROLLARY 12.4. *The only fixed points of the action of φ on the compactification of Teichmüller space $\overline{\mathcal{T}(M)}$ are $[\mathcal{F}^u, \mu^u]$ and $[\mathcal{F}^s, \mu^s]$.*

THEOREM 12.5 (Uniqueness of pseudo-Anosovs). *Two homotopic pseudo-Anosov diffeomorphisms are conjugate by a diffeomorphism isotopic to the identity.*

12.2   THE PERRON–FROBENIUS THEOREM AND MARKOV PARTITIONS

THEOREM 12.6 (Perron–Frobenius theorem). *Let $A = (a_{ij})$ be an $n \times n$ matrix with nonnegative entries. We denote by $a_{ij}^{(k)}$ the coefficients of A^k, the kth power of A. If there exists $\ell \geq 1$ such that all the coefficients $a_{ij}^{(\ell)}$ of A^ℓ are strictly positive, we have the following properties.*

1. *The matrix A admits an eigenvalue $\lambda > 0$ that is strictly greater than the absolute value of every other eigenvalue.*

2. *There exists an $x = (x_1, \ldots, x_n) \in \mathbb{R}^n$, with each x_i positive, that is an eigenvector for A with eigenvalue λ:*

$$\lambda x_i = \sum_{j=1}^{n} a_{ij} x_j , \quad i = 1, \ldots, n.$$

3. *The eigenspace of A associated to λ is one-dimensional.*

4. If $y = (y_1, \ldots, y_n) \in \mathbb{R}^n$ is an eigenvector with eigenvalue λ for A^t, that is,

$$\lambda y_j = \sum_{i=1}^{n} y_i a_{ij}, \quad j = 1, \ldots, n,$$

then all the y_j are positive. If y is normalized by

$$\langle x, y \rangle = \sum_{i=1}^{n} x_i y_i = 1,$$

we have

$$\lim_{k \to \infty} \frac{A^k}{\lambda^k} = \langle \ , y \rangle x;$$

that is,

$$\lim_{k \to \infty} \frac{a_{ij}^{(k)}}{\lambda^k} = x_i y_j.$$

For more details, see [Gan98, Chapter 13] or [Kar66, Appendix].

Markov partitions and the Perron–Frobenius theorem. We are going to consider a (pre-)Markov partition $\mathcal{R} = \{R_1, \ldots, R_N\}$ for φ (see Exposé 9). We set

$$\begin{aligned} x_i &= \mu^s(\mathcal{F}^u\text{-fiber of } R_i) \quad \text{and} \\ y_i &= \mu^u(\mathcal{F}^s\text{-fiber of } R_i). \end{aligned}$$

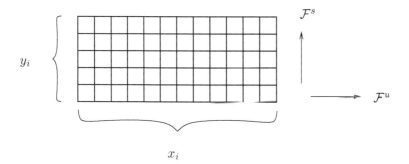

Figure 12.1.

We made the hypothesis that $\mu^s \otimes \mu^u(M) = 1$; this is equivalent to the condition

$$\sum_{i=1}^{N} x_i y_i = 1.$$

Let $A = (a_{ij})$ be the incidence matrix of $\mathcal{R}$ for φ, so

$$a_{ij} = (\text{number of times that } \varphi(\text{int}(R_i)) \text{ traverses int}(R_j)).$$

We have

$$\lambda x_i = \sum_{j=1}^{N} a_{ij} x_j, \text{ and} \tag{12.1}$$

$$\lambda y_i = \sum_{j=1}^{N} y_j a_{ji}. \tag{12.2}$$

We saw at the end of Section 10.4 that λ is in fact the greatest eigenvalue for A. Moreover, in Lemma 10.21, we showed that there exists an integer $\ell > 0$ such that the matrix A^ℓ has only positive entries. We can thus apply the Perron–Frobenius theorem, which gives us the following.

LEMMA 12.7. *With the notations introduced above, we have*

$$\lim_{k \to \infty} \frac{a_{ij}^{(k)}}{\lambda^k} = x_i y_j.$$

12.3 UNIQUE ERGODICITY

Let ν be an invariant measure for $\mathcal{F}^u$. Our goal is to show that ν differs from μ by a constant.

Since $\mathcal{F}^u$ does not have any leaves that are closed in $M - \text{Sing}(\mathcal{F}^u)$, the measure ν does not have any atoms. For each R_i, we choose an $\mathcal{F}^s$-fiber of R_i that passes through a point $p_i \in R_i$; denote this fiber by F_i (Figure 12.2).

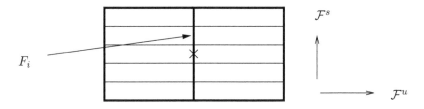

Figure 12.2.

LEMMA 12.8. *Let ν be an invariant measure for $\mathcal{F}^u$. There exists a constant C such that $\nu(F_i) = C\mu^u(F_i)$, $i = 1, \ldots, N$.*

Proof. We can suppose $\nu \geq 0$. Let $i \in \{1, \ldots, N\}$ be fixed. For any $k > 0$, we have

$$F_i = \left[\bigcup_{j=1}^{N} [\varphi^k(\text{int}(R_j)) \cap F_i] \right] \cup \{\text{a finite number of points}\}.$$

Now, since ν does not have atoms and since the $\text{int}(R_j)$ are pairwise disjoint, we have

$$\nu(F_i) = \sum_{j=1}^{N} \nu(\varphi^k(\text{int}(R_j)) \cap F_i).$$

In addition, by the properties of Markov partitions, $\varphi^k(\text{int}(R_j)) \cap F_i$ is a disjoint union of a certain number of intervals that, outside of their endpoints, all are the image of some $\varphi^k(F_j)$ under holonomy of $\varphi^k(F_j)$. Moreover, the number of these intervals is equal to the number of times that $\varphi^k(\text{int}(R_j))$ traverses $\text{int}(R_i)$, that is to say $a_{ji}^{(k)}$, and so

$$\nu(F_i) = \sum_{j=1}^{N} a_{ji}^{(k)} \nu(\varphi^k(F_j)). \tag{12.3}$$

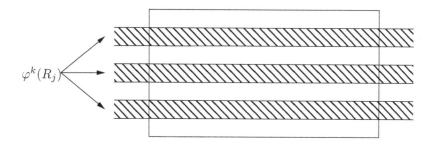

Figure 12.3.

We obtain in particular that $\nu(F_i) \geq a_{ji}^{(k)} \nu(\varphi^k(F_j))$. Since $a_{ji}^{(k)}/\lambda^k$ tends to the nonzero finite limit $x_j y_i$ as k tends to infinity, it follows that $\lambda^k \nu(\varphi^k(F_j))$ stays bounded as k tends to infinity, and in particular

$$\lim_{k \to \infty} \left(\frac{a_{ji}^{(k)}}{\lambda^k} - x_j y_i \right) \lambda^k \nu(\varphi^k(F_j)) = 0.$$

Combining this with the equality 12.3, we obtain

$$\nu(F_i) = \lim_{k \to \infty} \sum_{j=1}^{N} x_j y_i \lambda^k \nu(\varphi^k(F_j)) = y_i \lim_{k \to \infty} \left(\sum_{j=1}^{N} \lambda^k x_j \nu(\varphi^k(F_j)) \right).$$

This completes the proof of the lemma, since $y_i = \mu^u(F_i)$, and

$$\lim_{k \to \infty} \sum_{j=1}^{N} \lambda^k x_j \nu(\varphi^k(F_j))$$

is a constant that is independent of i.   □

We now complete the proof of Theorem 12.1. For each $m \geq 0$, we consider the Markov partition $\{\Gamma^k_{m,i,j}\}$ given by the closures of the connected components of $\varphi^m(\text{int}(R_i)) \cap \text{int}(R_j)$. Lemma 12.8 gives that, for fixed m, there exists a constant C_m such that

$$\forall \; k, i, j, \quad \nu(F_j \cap \Gamma^k_{m,i,j}) = C_m \mu^u(F_j \cap \Gamma^k_{m,i,j}).$$

If we fix j in the above equalities, and if we sum over k and i, we obtain $\nu(F_j) = C_m \mu^u(F_j)$; thus C_m is independent of m. We have thus shown the existence of a constant C such that

$$\forall \; m, k, i, j, \quad \nu(F_j \cap \Gamma^k_{m,i,j}) = C\mu^u(F_j \cap \Gamma^k_{m,i,j}). \tag{12.4}$$

For fixed m, the $F_j \cap \Gamma^k_{m,i,j}$ give a covering of F_j by intervals that only intersect at their endpoints. Moreover, each $F_j \cap \Gamma^k_{m,i,j}$ is included in one $F_j \cap \Gamma^{k'}_{m,i',j}$ and the diameter of $F_j \cap \Gamma^k_{m,i,j}$ tends to zero as m tends to infinity. From these properties and the fact that ν and μ^u have no atomic masses, the equalities 12.4 imply

$$\nu|_{F_j} = C\mu^u|_{F_j}, \quad j = 1, \ldots, N.$$

It follows that $\nu = C\mu^u$ since each leaf intersects each F_j.

12.4   THE ACTION OF PSEUDO-ANOSOVS ON $\mathcal{PMF}$

We begin with some generalities about quasitransverse curves. We consider an orientable surface N without boundary (compact or not) and $\mathcal{F}$ a measured foliation on N. We will say that an immersed (closed) curve is *quasitransverse* to $\mathcal{F}$ if it is an immersion $S^1 \xrightarrow{j} N$ with the following properties:

(i) It is a limit of embeddings.

(ii) It only has a finite number of double points and moreover, these double points are points of $\text{Sing}(\mathcal{F})$.

(iii) It is quasitransverse to $\mathcal{F}$ (cf. Section 5.2).

PROPOSITION 12.9. *Let $\alpha : [0, 1] \to N$ be a path that is quasitransverse to $\mathcal{F}$, with $\alpha(0) = \alpha(1)$ and with no other double points. Suppose that α leaves and arrives transversely to $\mathcal{F}$. Then the closed curve defined by α is not nullhomotopic.*

Proof. We denote by x_0 the origin/endpoint of α. We consider the case where x_0 is a regular point of $\mathcal{F}$. The situation in a neighborhood of x_0 is one of the two indicated in Figure 12.5.

In the first case, we have a quasitransverse curve which, by Proposition 5.6, cannot be nullhomotopic. In the second case, we construct a curve homotopic to

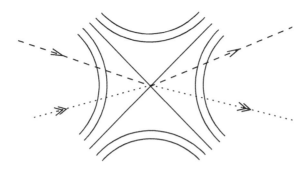

Figure 12.4.

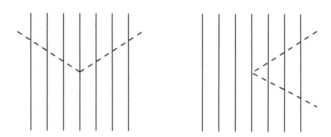

Figure 12.5.

α with a piece of α and a small piece of a leaf; this curve cannot be nullhomotopic by Proposition 5.6.

Considering the case $x_0 \in \mathrm{Sing}(\mathcal{F})$, we have three possible configurations (Figure 12.6). The first case gives us an embedded quasitransverse curve. In the second and third cases, we reduce to the case where x_0 is nonsingular by making the modifications shown in Figure 12.7.

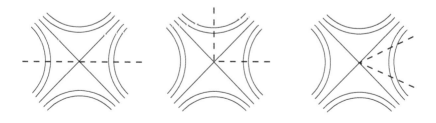

Figure 12.6.

We obtain, again by Proposition 5.6, that α is not nullhomotopic. $\square$

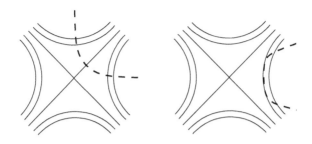

Figure 12.7.

COROLLARY 12.10. *If N is simply connected, then every immersion $\mathbb{R} \to N$ that is quasitransverse to $\mathcal{F}$ is simple.*

Proof. If this immersion has double points, we can find a path quasitransverse to $\mathcal{F}$ as in the hypothesis of Proposition 12.9 and that is nullhomotopic since N is simply connected. This contradicts Proposition 12.9. □

PROPOSITION 12.11. *Suppose that $N \cong S^1 \times \mathbb{R}$, and let α be an immersed curve that is quasitransverse to $\mathcal{F}$ and homotopic to the core of the cylinder $S^1 \times \{0\}$. Then α is a simple curve.*

Proof. Let $\mathbb{R}^2 = \tilde{N} \xrightarrow{p} N \cong S^1 \times \mathbb{R}$ be the universal covering of N. We think of α as a map $\mathbb{R} \xrightarrow{\alpha} N$ that is $\mathbb{Z}$-periodic. Let $\tilde{\alpha} : \mathbb{R} \to \tilde{N}$ be a lift of α. Since α is homotopic to the core of the cylinder, we can see that $p^{-1}(\alpha(0)) = \{\tilde{\alpha}(n) | n \in \mathbb{Z}\}$, and we see that $p^{-1}(\alpha) = \tilde{\alpha}(\mathbb{R})$. By Corollary 12.10, $\tilde{\alpha}(\mathbb{R}) = p^{-1}(\alpha)$ is simple, from which it follows that α is simple, since $p^{-1}(\alpha) \xrightarrow{p} \alpha$ is a covering. □

In what follows, we consider a pseudo-Anosov diffeomorphism; we denote by $(\mathcal{F}^s, \mu^s)$ and $(\mathcal{F}^u, \mu^u)$ its invariant foliations and by $\lambda > 1$ its dilatation coefficient.

LEMMA 12.12. *Let γ be an embedded curve in M that is not nullhomotopic. We can find immersed curves γ^s and γ^u that are quasitransverse to $\mathcal{F}^s$ and $\mathcal{F}^u$ and that are homotopic to γ.*

Proof. Recall that $\mathcal{F}^s$ has no connections between singularities. By Proposition 5.9, we can find a foliation $\mathcal{F}_1^s$ equivalent to $\mathcal{F}^s$ and an embedded curve γ_1^s transverse to $\mathcal{F}_1^s$ and isotopic to γ. When we recover $\mathcal{F}^s$ by blowing down the connections between the singularities, we transform the curve γ_1^s into the desired curve γ^s. □

We endow M with the metric $ds^2 = (d\mu^s)^2 + (d\mu^u)^2$, which is flat outside of the singularities. Below, when we talk about angles being small, this will only make sense away from the singularities. We remark that for this metric, the foliations $\mathcal{F}^s$ and $\mathcal{F}^u$ are orthogonal at all (regular) points.

LEMMA 12.13. *Let α be an immersed curve quasitransverse to $\mathcal{F}^s$ (resp. $\mathcal{F}^u$). The angle of $\varphi^n(\alpha)$ with $\mathcal{F}^u$ (resp. of $\varphi^{-n}(\alpha)$ with $\mathcal{F}^s$) tends to zero as n tends to infinity.*

The proof of this lemma is left to the reader.

PROPOSITION 12.14. *We consider in $\widetilde{M}$, the universal cover of M, the two induced foliations $(\widetilde{\mathcal{F}}^s, \widetilde{\mu}^s)$, $(\widetilde{\mathcal{F}}^u, \widetilde{\mu}^u)$ and the flat metric $d\tilde{s}^2 = (d\widetilde{\mu}^s)^2 + (d\widetilde{\mu}^u)^2$. Let γ be a simple arc that is quasitransverse to $\widetilde{\mathcal{F}}^u$ and whose angle with $\widetilde{\mathcal{F}}^s$ is less than $\pi/4$, and let δ be a simple arc that is quasitransverse to $\widetilde{\mathcal{F}}^s$ and whose angle with $\widetilde{\mathcal{F}}^u$ is less than $\pi/4$. Then $\gamma \cup \delta$ cannot be a simple closed curve.*

Proof. We suppose that $\gamma \cup \delta$ is a simple closed curve; as $\widetilde{M} \cong \mathbb{R}^2$, it bounds a disk Δ. If δ passes through a singularity s_0, then a local isotopy makes δ coincide with arcs of the separatrices of $\widetilde{\mathcal{F}}^u$ in a neighborhood of s_0; we can perform this operation while preserving the conditions on angles and keeping $\gamma \cup \delta$ embedded. Now, by the angle condition on δ, the field of tangent vectors to $\widetilde{\mathcal{F}}^u$ along δ can be turned without ambiguity until it becomes tangent to δ; the angle condition on γ allows us to extend this field to a field that is quasitransverse to γ, and that coincides with $\widetilde{\mathcal{F}}^u$ in a neighborhood of the singularities and outside of a neighborhood of δ. The new foliation has only permissible singularities, which gives us a contradiction with the Euler–Poincaré formula (Theorem 5.1). □

COROLLARY 12.15. *Let α and β be two immersed curves that are quasitransverse to $\mathcal{F}^s$ and $\mathcal{F}^u$, respectively, and such that the angle of α with $\mathcal{F}^u$ (resp. β with $\mathcal{F}^s$) is less than $\pi/4$. Two lifts $\tilde{\alpha}$ and $\tilde{\beta}$ in $\widetilde{M}$ intersect in at most one point.*

Proof. By Corollary 12.10, the immersions $\tilde{\alpha}$ and $\tilde{\beta}$ are embeddings. If card($\tilde{\alpha} \cap \tilde{\beta}$) ≥ 2, we can find a disk Δ with $\partial\Delta = \gamma \cup \delta$ where $\gamma \subset \tilde{\beta}$ and $\delta \subset \tilde{\alpha}$, which is absurd by the preceding proposition. □

Let α and β be two simple curves in M that are not nullhomotopic. We denote by α' (resp. β') an immersion quasitransverse to $\mathcal{F}^s$ (resp. $\mathcal{F}^u$) and homotopic to α (resp. β). We denote by $P(\alpha')$ (resp. $P(\beta')$) the number of times that α' (resp. β') passes through a singularity. We denote by Int(α', β') the number of points of intersection of α' and β' counted with multiplicity in the following manner: let $\{p_1, \ldots, p_k\} = \alpha' \cap \beta'$ (with $p_i \neq p_j$ when $i \neq j$); we assign to p_i the multiplicity

$$m_i = \text{(number of times } \beta' \text{ passes through } p_i) \times \text{(number of times } \alpha' \text{ passes through } p_i);$$

so, by definition,

$$\text{Int}(\alpha', \beta') = \sum_{i=1}^{k} m_i.$$

PROPOSITION 12.16. *For all $n, k \geq 0$ large enough, we have*

$$\left| i(\varphi^n(\alpha), \varphi^{-k}(\beta)) - \mathrm{Int}(\varphi^n(\alpha'), \varphi^{-k}(\beta')) \right| \leq P(\alpha')P(\beta').$$

Proof. We start by noting that

$$P(\varphi^n(\alpha')) = P(\alpha')$$

and

$$P(\varphi^{-k}(\beta')) = P(\beta').$$

By Lemma 12.13, the angle of $\varphi^n(\alpha')$ with $\mathcal{F}^u$ (resp. $\varphi^{-k}(\beta')$ with $\mathcal{F}^s$) is less than $\pi/4$ for n (resp. k) sufficiently large. It thus suffices to show that

$$|i(\alpha, \beta) - \mathrm{Int}(\alpha', \beta')| \leq P(\alpha')P(\beta')$$

if the angle of α' with $\mathcal{F}^u$ (resp. β' with $\mathcal{F}^s$) is less than $\pi/4$. Actually, since $\mathrm{Int}(\varphi^n(\alpha'), \varphi^{-k}(\beta'))$ is greater than $i(\varphi^n(\alpha), \varphi^{-k}(\beta))$, we need to show only that

$$\mathrm{Int}(\varphi^n(\alpha'), \varphi^{-k}(\beta')) \leq i(\varphi^n(\alpha), \varphi^{-k}(\beta)) + P(\alpha')P(\beta').$$

We denote by $\overline{M} \xrightarrow{\bar{p}} M$ the covering of M where $\bar{p}_\star(\pi_1(\overline{M}))$ is the cyclic group generated by α'. As M is orientable, we have $\overline{M} \cong S^1 \times \mathbb{R}$. Let $\bar{\alpha}'$ be a closed lift of α' in $\overline{M}$. Since $\bar{\alpha}'$ is quasitransverse to $\bar{\mathcal{F}}^s = \bar{p}^{-1}(\mathcal{F}^s)$ and it is homotopic to the core of the cylinder, $\bar{\alpha}'$ is in fact a simple curve by Proposition 12.11.

We consider β' as a $\mathbb{Z}$-periodic map $\mathbb{R} \xrightarrow{\beta'} M$. A lift of β' in $\overline{M}$ is by definition a lift $\bar{\beta}' : \mathbb{R} \to \overline{M}$ of the map $\beta' : \mathbb{R} \to M$.

We show that a lift $\bar{\beta}'$ of β' intersects $\bar{\alpha}'$ in at most one point (a point of $\bar{\alpha}' \cap \bar{\beta}'$ is counted with multiplicity if $\bar{\beta}'$ passes through this point multiple times, so a single point of intersection means that $\mathrm{card}(\bar{\alpha}' \cap \bar{\beta}') = 1$ and $\bar{\beta}'$ passes through $\bar{\alpha}' \cap \bar{\beta}'$ only once). To see this, suppose that $\bar{\beta}'(a) \in \bar{\alpha}'$ and $\bar{\beta}'(b) \in \bar{\alpha}'$ with $a \neq b$. As $\pi_1(\overline{M})$ is generated by $\bar{\alpha}'$, one can find a path $\bar{\gamma} : [a, b] \to \overline{M}$ such that $\bar{\gamma}([a, b]) \subset \bar{\alpha}', \bar{\gamma}(a) = \bar{\beta}'(a), \bar{\gamma}(b) = \bar{\beta}'(b)$ and such that $\bar{\beta}'|_{[a,b]}$ is homotopic to $\bar{\gamma}|_{[a,b]}$ with endpoints fixed. If we go up to the universal cover $\widetilde{M}$, we find a lift $\tilde{\alpha}'$ of α' and a lift $\tilde{\beta}'$ of β' in $\widetilde{M}$ such that $\mathrm{card}(\tilde{\alpha}' \cap \tilde{\beta}') \geq 2$; however, this is impossible by Corollary 12.15 (here is where we use the assumption on the angles).

It follows that $\mathrm{Int}(\alpha', \beta')$ is equal to the number of lifts of β' in $\overline{M}$ that intersect $\bar{\alpha}'$. The lifts $\bar{\beta}'$ that intersect $\bar{\alpha}'$ are partitioned into two categories. The first category consists of lifts that lie on a single side of $\bar{\alpha}'$, and the second category consists of lifts that join the two infinities of the cylinder $\overline{M}$ (Figure 12.8).

Since α' and β' are transverse outside of the singularities of $\mathcal{F}^s$ (or $\mathcal{F}^u$), it is easy to see that the point of contact for the first category is a singularity. We conclude that the number of lifts of β' that intersect $\bar{\alpha}'$ and that are in the first category is at most $P(\alpha')P(\beta')$. The reader will easily check that the number of

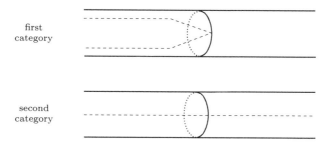

first
category

second
category

Figure 12.8.

lifts of β' that intersect $\bar{\alpha}'$ and that are in the second category is in fact exactly $i(\alpha, \beta)$. □

We now set about proving Theorem 12.2. We consider a Markov partition $\mathcal{R} = \{R_1, \ldots, R_N\}$ for φ as in Section 12.2. We can, by small perturbations, suppose that α' and β' are transverse to

$$\bigcup_{i=1}^{N} \partial_{\mathcal{F}^u} R_i \qquad \text{and} \qquad \bigcup_{i=1}^{N} \partial_{\mathcal{F}^s} R_i,$$

respectively. As $\varphi^{-1}(\cup \partial_{\mathcal{F}^u} R_i) \subset \cup \partial_{\mathcal{F}^u} R_i$, the curve $\varphi^{\ell}(\alpha')$ is also transverse to $\cup \partial_{\mathcal{F}^u} R_i$ for $\ell \geq 0$; in the same way $\varphi^{-\ell}(\beta')$ is transverse to $\cup \partial_{\mathcal{F}^s} R_i$.

For $\ell \geq 0$, we denote by $\bar{\alpha}_i^{\ell}$ the number of connected components of the pre-image of R_i under a parametrization of $\varphi^{\ell}(\alpha')$. The image of any such component will be called a *passage* of $\varphi^{\ell}(\alpha')$ in R_i. We say that a passage is *good* if it does not meet $\partial_{\mathcal{F}^u} R_i$; otherwise we say that it is *bad*. We denote by α_i^{ℓ} the number of good passages of $\varphi^{\ell}(\alpha')$ in R_i.

good passage

bad passages

$\partial_{\mathcal{F}^u} R_i$

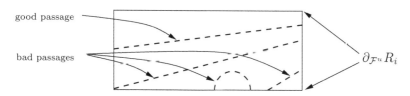

Figure 12.9.

We remark that $\bar{\alpha}_i^{\ell} - \alpha_i^{\ell}$ (the number of bad passages) is bounded above by the number of times (with multiplicity) that $\varphi^{\ell}(\alpha')$ intersects $\partial_{\mathcal{F}^u} R_i$. As $\varphi^{-\ell}(\cup \partial_{\mathcal{F}^u} R_i) \subset \cup \partial_{\mathcal{F}^u} R_i$, if C_1 denotes the number of times that α' intersects $\cup \partial_{\mathcal{F}^u} R_i$, we find that

$$\alpha_i^{\ell} \leq \bar{\alpha}_i^{\ell} \leq \alpha_i^{\ell} + C_1.$$

In the same manner, we define $\bar{\beta}_i^\ell$ and β_i^ℓ by replacing $\varphi^\ell(\alpha')$ by $\varphi^{-\ell}(\beta')$ and $\mathcal{F}^u$ by $\mathcal{F}^s$. We also find a constant C_2 such that $\beta_i^\ell \le \bar{\beta}_i^\ell \le \beta_i^\ell + C_2$, for all $i = 1, \ldots, N$, and for all $\ell \ge 0$. We set $C = \max(C_1, C_2)$.

Since $\varphi^{-n}(\cup \partial_{\mathcal{F}^u} R_i) \subset \cup \partial_{\mathcal{F}^u} R_i$ and $\varphi^n(\cup \partial_{\mathcal{F}^s} R_i) \subset \cup \partial_{\mathcal{F}^s} R_i$ for $n \ge 0$, it is easy to see that if P is a good passage of $\varphi^\ell(\alpha')$ in R_i, then $\varphi^n(P) \cap R_j$ is composed of $a_{ij}^{(n)}$ good passages of $\varphi^{\ell+n}(\alpha')$, where $A = (a_{ij})$ is the incidence matrix associated to the Markov partition and $A^n = \left(a_{ij}^{(n)} \right)$. On the other hand, if P' is an arbitrary passage of $\varphi^\ell(\alpha')$ in R_i, then $\varphi^n(P') \cap R_j$ is composed of at most $a_{ij}^{(n)}$ passages of $\varphi^{\ell+n}(\alpha')$ in R_j (here we use the fact that α' is quasitransverse to $\mathcal{F}^s$). We therefore have the following inequalities:

$$\sum_{i=1}^N \alpha_i^\ell a_{ij}^{(n)} \le \alpha_j^{\ell+n} \le \bar{\alpha}_j^{\ell+n} \le \sum_{i=1}^N \bar{\alpha}_i^\ell a_{ij}^{(n)} \le \sum_{i=1}^N (\alpha_i^\ell + C) a_{ij}^{(n)}.$$

We recall that $x_i = \mu^s(\mathcal{F}^u\text{-fiber of } R_i)$ and $y_i = \mu^u(\mathcal{F}^s\text{-fiber of } R_i)$.

CLAIM 12.17. *We have*

$$\lim_{\ell \to \infty} \sum_{i=1}^N \frac{x_i \alpha_i^\ell}{\lambda^\ell} = \mathrm{I}(\mathcal{F}^s, \mu^s; \alpha) \tag{12.5}$$

$$\lim_{\ell \to \infty} \sum_{j=1}^N \frac{y_j \beta_j^\ell}{\lambda^\ell} = \mathrm{I}(\mathcal{F}^u, \mu^u; \beta) \tag{12.6}$$

Proof. The quantity $\mathrm{I}(\mathcal{F}^s, \mu^s; \varphi^\ell(\alpha)) = \lambda^\ell \mathrm{I}(\mathcal{F}^s, \mu^s; \alpha)$ is nothing other than the μ^s-length of $\varphi^\ell(\alpha')$, since $\varphi^\ell(\alpha')$ is quasitransverse to $\mathcal{F}^s$ and is homotopic to $\varphi^\ell(\alpha)$. Also

$$\sum_{i=1}^N x_i \alpha_i^\ell \le \mu^s(\varphi^\ell(\alpha')) \le \sum_{i=1}^N x_i \bar{\alpha}_i^\ell \le \left(\sum_{i=1}^N x_i \alpha_i^\ell \right) + C \sum_{i=1}^N x_i,$$

which implies

$$\sum_{i=1}^N x_i \alpha_i^\ell \le \lambda^\ell \mathrm{I}(\mathcal{F}^s, \mu^s; \alpha) \le \left(\sum_{i=1}^N x_i \alpha_i^\ell \right) + C \sum_{i=1}^N x_i.$$

Equation 12.5 follows easily from this inequality, and equation 12.6 is obtained by interchanging the roles. $\qquad \square$

We remark that we have

$$\sum_{j=1}^N \alpha_j^{n+\ell} \beta_j^\ell \le \mathrm{Int}(\varphi^{n+\ell}(\alpha'), \varphi^{-\ell}(\beta')).$$

By Lemma 12.13 and Corollary 12.15, for ℓ large enough and $n \geq 0$, we have

$$\text{Int}(\varphi^{n+\ell}(\alpha'), \varphi^{-\ell}(\beta')) \leq \sum_{j=1}^{N} \bar{\alpha}^{n+\ell} \bar{\beta}_j^\ell,$$

whence by the inequalities written above we have

$$\sum_{i,j} \alpha_i^\ell a_{ij}^{(n)} \beta_j^\ell \leq \text{Int}(\varphi^{n+\ell}(\alpha'), \varphi^{-\ell}(\beta')) \leq \sum_{i,j} (\alpha_i^\ell + C) a_{ij}^{(n)} (\beta_j^\ell + C).$$

By Proposition 12.16, for ℓ large enough and $n \geq 0$, we have

$$|\text{Int}(\varphi^{n+\ell}(\alpha'), \varphi^{-\ell}(\beta')) - i(\varphi^{n+\ell}(\alpha), \varphi^{-\ell}(\beta))| \leq P(\alpha')P(\beta').$$

We also have of course

$$i(\varphi^{n+\ell}(\alpha), \varphi^{-\ell}(\beta)) = i(\varphi^{n+2\ell}(\alpha), \beta).$$

Combining the preceding inequalities, for ℓ large we obtain

$$\left(\sum_{i,j} \alpha_i^\ell a_{ij}^{(n)} \beta_j^\ell \right) - P(\alpha')P(\beta')$$
$$\leq i(\varphi^{n+2\ell}(\alpha), \beta)$$
$$\leq \sum_{i,j} (\alpha_i^\ell + C) a_{ij}^{(n)} (\beta_j^\ell + C) + P(\alpha')P(\beta').$$

Dividing these inequalities by $\lambda^{n+2\ell}$ and applying Lemma 12.7, and then letting n tend to infinity and making (the fixed number) ℓ large enough, we find

$$\sum_{i,j} \frac{\alpha_i^\ell x_i y_j \beta_j^\ell}{\lambda^{2\ell}} \leq \liminf_{k \to \infty} \frac{i(\varphi^k(\alpha), \beta)}{\lambda^k}$$
$$\leq \limsup_{k \to \infty} \frac{i(\varphi^k(\alpha), \beta)}{\lambda^k}$$
$$\leq \sum_{i,j} \frac{(\alpha_i^\ell + C) x_i y_j (\beta_j^\ell + C)}{\lambda^{2\ell}}.$$

By Claim 12.17, if we let ℓ tend to infinity, we obtain

$$\lim_{k \to \infty} \frac{i(\varphi^k(\alpha), \beta)}{\lambda^k} = \text{I}(\mathcal{F}^s, \mu^s; \alpha) \text{I}(\mathcal{F}^u, \mu^u; \beta).$$

This completes the proof of Theorem 12.2.

Corollary 12.3 is an immediate consequence of Theorem 12.2.

Proof of Corollary 12.4. As we have seen (Theorem 9.16), the action of φ cannot have any fixed points in Teichmüller space $\mathcal{T}(M)$, and moreover, a nontrivial power of φ cannot preserve an isotopy class of simple curves. Supposing then that we have a fixed point from the action of φ on $\overline{\mathcal{T}(M)}$, this fixed point is an element $[\mathcal{F}, \mu]$ of $\mathcal{PMF}(M)$. In other words, there exists $\rho > 0$ such that $\varphi(\mathcal{F}, \mu) \overset{m}{\sim} (\mathcal{F}, \rho\mu)$. It follows that $\mathcal{F}$ is arational, because otherwise a nontrivial power of φ preserves an isotopy class of curves. Moreover, ρ is different from 1, because otherwise φ would be isotopic to a periodic diffeomorphism (see Section 9.4). We suppose that $\rho > 1$; the case $\rho < 1$ is treated in the same manner. We can then, by Section 9.5, isotope φ to a pseudo-Anosov diffeomorphism φ' that admits $(\mathcal{F}, \mu)$ for an unstable foliation. Corollary 12.3 applied to φ and to φ' gives the following for all $\alpha \in \mathcal{S}(M)$:

$$\lim_{n \to \infty} [\varphi^n(\alpha)] = [\mathcal{F}^u, \mu^u] \text{ in } \mathcal{PMF}(M),$$
$$\lim_{n \to \infty} [(\varphi')^n(\alpha)] = [\mathcal{F}, \mu] \text{ in } \mathcal{PMF}(M).$$

As φ and φ' are isotopic, we obtain $[\mathcal{F}^u, \mu^u] = [\mathcal{F}, \mu]$. The case $\rho < 1$ would give $[\mathcal{F}^s, \mu^s] = [\mathcal{F}, \mu]$.   □

12.5   UNIQUENESS OF PSEUDO-ANOSOV MAPS

Our final task is to prove Theorem 12.5, that homotopic pseudo-Anosov maps are conjugate. We begin by proving two lemmas.

LEMMA 12.18. *Let M be a closed, orientable surface of genus $g \geq 2$ and let φ be a diffeomorphism of M isotopic to the identity. If φ is periodic, then φ is the identity.*

Proof. We have seen (in the remark at the end of Section 9.4) that the uniformization theorem implies that φ is an isometry for some hyperbolic metric. Since φ is isotopic to the identity, φ is in fact equal to the identity (Theorem 3.19).   □

LEMMA 12.19. *Let $(\mathcal{F}^u, \mu^u)$ be an arational foliation of M and let φ be a diffeomorphism of M, isotopic to the identity, that preserves $(\mathcal{F}^u, \mu^u)$. Then φ is isotopic to the identity through diffeomorphisms that preserve $(\mathcal{F}^u, \mu^u)$.*

Proof. Lemma 9.7 says that φ is isotopic to a periodic diffeomorphism φ', through diffeomorphisms that preserve $(\mathcal{F}^u, \mu^u)$. The preceding lemma shows that φ' is the identity.   □

Let φ_1 and φ_2 be two isotopic pseudo-Anosov diffeomorphisms. Denote by $(\mathcal{F}_1^u, \mu_1^u)$ and $(\mathcal{F}_2^u, \mu_2^u)$ the unstable foliations of φ_1 and φ_2, and by $(\mathcal{F}_1^s, \mu_1^s)$ and $(\mathcal{F}_2^s, \mu_2^s)$ the stable foliations of φ_1 and φ_2. By Corollary 12.4 we have that $[\mathcal{F}_1^u, \mu_1^u] = [\mathcal{F}_2^u, \mu_2^u]$ in $P(\mathbb{R}_+^{\mathcal{S}})$.

Up to multiplying $(\mathcal{F}_1^u, \mu_1^u)$ by a positive nonzero constant, we can thus suppose that $(\mathcal{F}_1^u, \mu_1^u) = (\mathcal{F}_2^u, \mu_2^u)$ in $\mathcal{MF}$. Since these foliations do not have connections between singularities, there exists a diffeomorphism h isotopic to the identity such that $(\mathcal{F}_1^u, \mu_1^u) = h(\mathcal{F}_2^u, \mu_2^u)$ where the equality means here that the foliations in M are the same and the transverse measures are the same. Up to replacing φ_2 by $h\varphi_2 h^{-1}$, we are reduced to the case where φ_1 and φ_2 have the same unstable foliation $(\mathcal{F}^u, \mu^u)$. We also remark that the expansion constant $\lambda \ (> 1)$ is the same for φ_1 and φ_2; this follows, for example, from the fact that $\varphi_1(\mathcal{F}^u, \mu^u) = \varphi_2(\mathcal{F}^u, \mu^u)$ in $\mathcal{MF}$. It follows that $\varphi_2^{-1}\varphi_1$ preserves $(\mathcal{F}^u, \mu^u)$. By Lemma 12.19, $\varphi_2^{-1}\varphi_1$ is isotopic to the identity through diffeomorphisms that preserve $(\mathcal{F}^u, \mu^u)$; we denote by h_t one such isotopy. In particular, for every x in M, $\varphi_2^{-1}\varphi_1(x)$ and x are on the same $\mathcal{F}^u$-leaf. We denote by $[x, \varphi_2^{-1}\varphi_1(x)]$ the segment of the $\mathcal{F}^u$-leaf of x that joins x to $\varphi_2^{-1}\varphi_1(x)$.

CLAIM 12.20. *We have*

$$D = \sup\{\mu_2^s([x, \varphi_2^{-1}\varphi_1(x)]) \mid x \in M\} < \infty.$$

Proof. Let $U_1, \ldots, U_k$ be a covering of M by charts for the foliation $\mathcal{F}^u$. We denote by A the subset of $M \times M$ defined by the condition that $(x, y) \in A$ if there exists a plaque of $\mathcal{F}^u$ contained in one of the U_i and that contains x and y (in particular, since the "plaque" of a singular point is a single point, if $(x, y) \in A$ and x (or y) is a singular point of $\mathcal{F}^u$, then $x = y$). If $(x, y) \in A$, we denote by $[x, y]$ the segment of the plaque that contains x and y and that goes from x to y. The function $(x, y) \to \mu_2^s([x, y])$ is continuous on A. We consider then the isotopy h_t of $\varphi_2^{-1}\varphi_1$ to the identity, through homeomorphisms that preserve $\mathcal{F}^u$. We can find a $\delta > 0$ such that, if $|t - t'| < \delta$, then $(h_t(x), h_{t'}(x)) \in A$; by the compactness of M and what has been said above, we have

$$D_{t,t'} = \sup\{\mu_2^s([h_t(x), h_{t'}(x)]) \mid x \in M\} < \infty.$$

We then consider a sequence $t_0 = 0 < t_1 < \cdots < t_{n-1} < t_n = 1$ such that $t_{i+1} - t_i < \delta$. For all $x \in M$, we have

$$\mu_2^s([x, \varphi_2^{-1}\varphi_1(x)]) \leq \sum_{i=0}^{n-1} \mu_2^o([h_{t_i}(x), h_{t_{i+1}}(x)]);$$

from this we have

$$D \leq \sum_{i=0}^{n-1} D_{t_i, t_{i+1}} < \infty.$$

$\square$

CLAIM 12.21. *The sequence of homeomorphisms* $(\varphi_2^{-n}\varphi_1^n)_{n \geq 0}$ *converges uniformly.*

Proof. Let d be the metric obtained from $ds^2 = (d\mu_2^s)^2 + (d\mu_2^u)^2$. We remark that if x and y are on the same $\mathcal{F}^u$-leaf, and if $[x, y]$ denotes the segment of this leaf that goes from x to y, we have $d(x,y) \leq \mu_2^s([x,y])$ (we do not have equality in general, because, since the leaves demonstrate recurrence, two points can be close in M without the segment of the leaf that goes from one to the other being small). The uniform convergence of the sequence follows easily from the following inequality that we are going to establish:

$$\sup_{x \in M} d(\varphi_2^{-(n+1)}\varphi_1^{(n+1)}(x), \varphi_2^{-n}\varphi_1^n(x)) \leq \lambda^{-n} D.$$

We consider the $\mathcal{F}^u$-segment $[\varphi_2^{-1}\varphi_1(\varphi_1^n(x)), \varphi_1^n(x)]$; its measure is at most D. The image of this segment under φ_2^{-n} is nothing other than the $\mathcal{F}^u$-segment

$$[\varphi_2^{-(n+1)}\varphi_1^{(n+1)}(x), \varphi_2^{-n}\varphi_1^n(x)].$$

Considering the effect of φ_2^{-n} on μ_2^s, we have

$$
\begin{aligned}
\mu_2^s([\varphi_2^{-(n+1)}\varphi_1^{(n+1)}(x), \varphi_2^{-n}\varphi_1^n(x)]) &\leq \lambda^{-n}\mu_2^s([\varphi_2^{-1}\varphi_1(\varphi_1^n(x)), \varphi_1^n(x)]) \\
&\leq \lambda^{-n} D,
\end{aligned}
$$

from which we obtain

$$d\left(\varphi_2^{-(n+1)}\varphi_1^{(n+1)}(x), \varphi_2^{-n}\varphi_1^n(x)\right) \leq \lambda^{-n} D.$$

$\square$

We denote by h the uniform limit of $(\varphi_2^{-n}\varphi_1^n)_{n \geq 0}$. We remark that h is invertible; one shows this in the same manner that one shows that the sequence of inverses $(\varphi_1^{-n}\varphi_2^n)_{n \geq 0}$ converges uniformly. We also remark that h is isotopic to the identity since each $\varphi_1^{-n}\varphi_2^n$ is isotopic to the identity.

We then consider $h\varphi_1$; we have

$$
\begin{aligned}
h\varphi_1 &= \left(\lim_{n \to \infty} \text{unif } \varphi_2^{-n}\varphi_1^n\right)\varphi_1 \\
&= \lim_{n \to \infty} \text{unif}\left(\varphi_2^{-n}\varphi_1^{(n+1)}\right) \\
&= \lim_{n \to \infty} \text{unif}\left(\varphi_2\left(\varphi_2^{-(n+1)}\varphi_1^{(n+1)}\right)\right) \\
&= \varphi_2 \lim_{n \to \infty} \text{unif}\left(\varphi_2^{-(n+1)}\varphi_1^{(n+1)}\right) \\
&= \varphi_2 h,
\end{aligned}
$$

so $h\varphi_1 = \varphi_2 h$, which shows that h is a conjugation between φ_1 and φ_2.

To prove Theorem 12.5, it remains to check that h is differentiable. Before doing this, we must make the definition of pseudo-Anosov more precise; that is, we insist that, using a C^∞ chart in the neighborhood of a singularity, the foliations $\mathcal{F}^s$ and $\mathcal{F}^u$ are given by the absolute values of the real and imaginary parts of $\sqrt{z^{p-2}\,dz^2}$ $(p \geq 3)$.

LEMMA 12.22. *A conjugation between two pseudo-Anosov diffeomorphisms is automatically C^∞ differentiable.*

Outline of proof. Denote by h the conjugation, φ_1 and φ_2 the two pseudo-Anosov diffeomorphisms: $h\varphi_1 h^{-1} = \varphi_2$. The first thing we remark is that h sends the (un)stable foliation of φ_1 onto the (un)stable foliation of φ_2 (without talking for the moment about the transverse measure). This follows, for example, from the fact that the set

$$W_x^s(\varphi_i) = \left\{ y \in M \mid \lim_{n \to \infty} d(\varphi_i^n(x), \varphi_i^n(y)) = 0 \right\}$$

is either the leaf of $\mathcal{F}_i^s$ that passes through x (if this leaf does not emanate from a singularity) or it is the finite union of the leaves of $\mathcal{F}_i^s$ emanating from one singularity (if x is a singularity or if the leaf of $\mathcal{F}_i^s$ containing x emanates from a singularity). Moreover, as $\mathcal{F}_2^s$ (resp. $\mathcal{F}_2^u$) is uniquely ergodic (Theorem 12.1), h also sends μ_1^s (resp. μ_1^u) onto μ_2^s (resp. μ_2^u), up to dividing the measures by a suitable constant.

Considering then a regular point m for $\mathcal{F}_1^s$ and $\mathcal{F}_1^u$, its image $h(m)$ is a regular point for $\mathcal{F}_2^s$ and $\mathcal{F}_2^u$. We can find a smooth chart

$$(-\varepsilon, \varepsilon) \times (-\varepsilon, \varepsilon) \overset{\psi}{\to} M$$

such that $\psi(0) = m$ and that the foliation $(\mathcal{F}_1^s, \mu_1^s)$ (resp. $(\mathcal{F}_1^u, \mu_1^u)$) is defined in this chart by the 1-form dx (resp. dy). In the same way, we find one such chart around $h(m)$. When we read h in these charts, it takes the form of a homeomorphism of $(-\varepsilon, \varepsilon) \times (-\varepsilon, \varepsilon)$ on an open neighborhood of 0 in $\mathbb{R}^2$ that sends 0 to 0, the horizontals into the horizontals, the verticals into the verticals, and that preserves the spacing between two horizontals or two verticals. It is easy to see that h is the restriction to $(-\varepsilon, \varepsilon) \times (-\varepsilon, \varepsilon)$ of one of the following four linear maps of $\mathbb{R}^2$: identity, orthogonal symmetry with respect to the x-axis (resp. y-axis), reflection through the origin. It follows that h is C^∞ at every regular point.

We can make an analogous argument at a singular point. Recall that we have made precise the definition of a pseudo-Anosov. This precision implies that in suitable charts, h appears as a germ of a homeomorphism at $0 \in \mathbb{C}$ that preserves the absolute values of the real and imaginary parts of $\sqrt{z^{p-2}\,dz^2}$ ($p \geq 3$). The reader will verify that such germs, that preserve orientation, are rotations of angles $2k\pi/p$, and the germs that reverse orientation are given by symmetries with respect to the lines that contain the union of two separatrices. □

Remark. One may wonder—what are necessary and sufficient conditions so that a uniquely ergodic arational foliation is the stable foliation of a pseudo-Anosov diffeomorphism?

Exposé Thirteen

Constructing Pseudo-Anosov Diffeomorphisms

by François Laudenbach

13.1 GENERALIZED PSEUDO-ANOSOV DIFFEOMORPHISMS

A *measured foliation with spines* is a measured foliation for which we allow, in addition to the usual singularities (Exposé 5), those of Figure 13.1 (the figure represents two transverse measured foliations with spines).

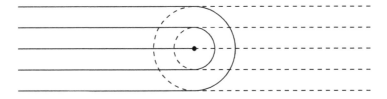

Figure 13.1. Two transverse measured foliations with spines

A *generalized pseudo-Anosov diffeomorphism* is a homeomorphism φ for which there exist two transverse measured foliations with spines, $(\mathcal{F}^s, \mu^s)$ and $(\mathcal{F}^u, \mu^u)$, and a scalar $\lambda > 1$, such that $\varphi(\mathcal{F}^s, \mu^s) = (\mathcal{F}^s, \lambda^{-1}\mu^s)$ and $\varphi(\mathcal{F}^u, \mu^u) = (\mathcal{F}^u, \lambda\mu^u)$.

The disk admits a measured foliation with spines that is transverse to the boundary (Figure 13.2). It is also possible that, for $\alpha \in \mathcal{S}$, one has $I(\mathcal{F}^s, \mu^s; \alpha) = 0$, even though $\mathcal{F}^s$ does not contain any cycles of leaves[1] (as φ contracts the μ^u-lengths, $\mathcal{F}^s$ does not have any connections between singularities, either). A generalized pseudo-Anosov diffeomorphism can fix an element of $\mathcal{S}$; in particular, it does not satisfy Proposition 9.21, which gives asymptotics for the growth rates of lengths of isotopy classes of curves (cf. footnote 1). Therefore, a generalized pseudo-Anosov can be isotopic to the identity; see the example on S^2 in Section 11.3.

Nevertheless, generalized pseudo-Anosovs are still useful because of the following remark. If one blows up the surface at the point of the spines, one obtains a pseudo-Anosov diffeomorphism of the surface with boundary. In particular,

---

[1]We construct one such example on T^2 by applying the construction of Section 13.3 to (α, β), where α is a "generator" of the torus and where β, isotopic to α, cuts $T^2 - \alpha$ into cells.

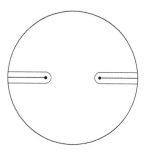

Figure 13.2. A measured foliation with spines that is transverse to the boundary

Poincaré recurrence still holds and one can construct a Markov partition. Thus a generalized pseudo-Anosov diffeomorphism is again a Bernoulli process.

13.2   A CONSTRUCTION BY RAMIFIED COVERS

Let $p : N \to M$ be a ramified cover of compact surfaces, and let $\Sigma \subset M$ be the locus of ramification. We suppose that above $M - \Sigma$ the covering is regular, with covering group G. Let φ be a generalized pseudo-Anosov of M. By isotopy of p, we can arrange for Σ to lie in the infinite set of periodic points of φ. Thus, up to replacing φ by one of its powers, we can suppose that $\varphi|_\Sigma$ is equal to the identity. The regular covering of p over $M - \Sigma$ is classified by an element of $H^1(M - \Sigma; G)$, a finite group on which φ acts. Up to again taking powers of φ, we can suppose that

$$\varphi^\star(p : N \to M) \approx (p : N \to M).$$

In other words, φ lifts to a diffeomorphism ψ. The local properties of φ are the same as those of ψ. Thus, ψ is a generalized pseudo-Anosov, with the same dilatation factor.

Now, every closed orientable surface N, of genus $g \geq 1$, is the total space of a ramified cover with two sheets over T^2, with a locus of ramification Σ satisfying

$$\operatorname{card}(\Sigma) = 2g - 2.$$

To see this, we puncture T^2 in $n = \operatorname{card}(\Sigma)$ points; this open manifold retracts onto a bouquet of $n+1$ circles $\varepsilon_1, \ldots, \varepsilon_{n+1}$. Say that each of $\varepsilon_1, \ldots, \varepsilon_{n-1}$ surrounds a hole and that ε_n and ε_{n+1} are "generators" of the torus. The last hole, denoted ∞, is surrounded homologically by $[\varepsilon_1] + \cdots + [\varepsilon_{n-1}]$. We construct the cover associated to the homomorphism $\pi_1(T^2 - \Sigma) \to \mathbb{Z}/2\mathbb{Z}$ that sends $\varepsilon_1, \ldots, \varepsilon_{n-1}$ to 1 and that takes any value on ε_n and ε_{n+1}. As n is even, the covering is nontrivial in a neighborhood of ∞. Thus the compactification gives a covering that is nontrivially ramified at each point of Σ.

Let φ be a (linear) Anosov map of T^2 that is the identity on Σ and that lifts to ψ in N. The stable foliation of ψ is transversely orientable and its singularities have four branches each. Thus ψ is pseudo-Anosov (not generalized) and the stable foliation is defined as a measured foliation by a closed 1-form ω^s; we have

$$\psi^\star \omega^s = \lambda \omega^s \quad (\lambda > 1).$$

In the same way, we have $\psi^\star \omega^u = \lambda^{-1}\omega^u$, where ω^u denotes the unstable foliation. We note that the two equalities together prohibit ψ from being differentiable at the singularities, but ψ can be approximated by a diffeomorphism ψ' that satisfies one of the equalities.

The disadvantage of this construction is that it is unmanageable on the level of calculation, for example, to compute the action of ψ on homology.

13.3   A CONSTRUCTION BY DEHN TWISTS

Suppose that the surface M is orientable and closed. In the case where there is boundary, one would start by filling the punctures and, at the end of the construction described below, one would puncture the corresponding number of singularities.

Flat structure on M.   Let α and β be two simple closed curves in M, with transverse intersection, satisfying the following condition:

Each component of $M - (\alpha \cup \beta)$ is an (open) cell.   (13.1)

The cellular decomposition induced on M by $\alpha \cup \beta$ admits a *dual* cellular decomposition: the co-vertices are the centers of the cells of $M - (\alpha \cup \beta)$. Each arc of $(\alpha \cup \beta) - (\alpha \cap \beta)$ is crossed by one co-edge. Each point x of $\alpha \cap \beta$ is the center of a co-cell, which is a square since α and β pass through x only once.

By "enlarging" α in the sense of Exposé 5, we construct a measured foliation $\mathcal{F}_\alpha$, that is transverse to β and transverse to the co-edges that meet α. We also arrange things so that the co-edges that do not meet α are contained in the leaves of $\mathcal{F}_\alpha$. Similarly, we construct $\mathcal{F}_\beta$ by suitably enlarging β. By isotopies in the interior of the co-cells, we can take $\mathcal{F}_\beta$ to be transverse to $\mathcal{F}_\alpha$. These foliations have their singularities at the co-vertices; in the complement, they define a flat structure. We understand these foliations better in the "unrolled" atlases shown in Figures 13.3 and 13.3.

If we unroll the co-cells along α, we obtain a band of $n = \text{card}(\alpha \cap \beta)$ squares, which we place in $\mathbb{R}^2$ in such a way that dy and dx, respectively, induce the measured foliations $\mathcal{F}_\alpha$ and $\mathcal{F}_\beta$.

In the same way, we construct the β-atlas by unrolling along β (respecting the orientation).

The transition maps are isometries of $\mathbb{R}^2$ with derivative $\pm I$, according to the sign of the intersection at the center of the co-cell (note that a transition

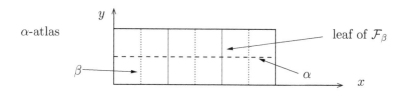

Figure 13.3.

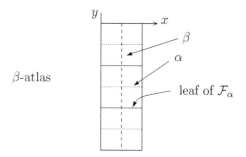

Figure 13.4.

map that preserves orientation cannot have the matrix $\left(\begin{smallmatrix} 1 & 0 \\ 0 & -1 \end{smallmatrix}\right)$ for its derivative). Consequently, relative to this atlas, the notion of a linear measured foliation is intrinsic. Moreover, the *slope* of the foliation is invariant under the transition maps. In M, the foliation is smooth except at the co-vertices (each point of the complement is interior to at least one chart). The co-vertices act like singularities (unless the corresponding cell is a square). The number of separatrices of the foliation at a vertex is half the number of sides of the corresponding cell (see Figure 13.3).

Remark. If all the points of intersection of α and β are of the same sign, then the transition maps have $+I$ for their derivative; thus, the orientation of the foliations is invariant under the transition maps. In other words, in this case, all linear foliations are orientable. Moreover, the atlas defines a function $M \to T^2$ which is an n-fold covering ramified at one point; but the covering is not regular and one cannot control the singular fiber.

Affine homeomorphisms. A homeomorphism φ is said to be *affine* if it leaves invariant the set of the co-vertices and if the image of a straight line of the flat structure is a straight line. Let $A(M)$ be the group of affine homeomorphisms.

The derivative of φ, modulo $\pm I$, is independent of the atlas used and independent of the point where one calculates it. We thus have a derivation homomorphism:

$$D : A(M) \to \mathrm{GL}(2, \mathbb{R})/\pm I.$$

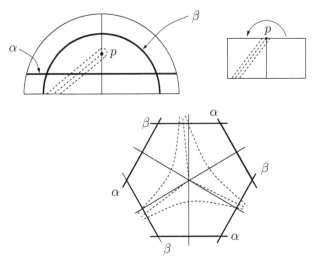

Figure 13.5.

For example, the positive Dehn twists along α and β admit affine representatives which, in the α- and β-atlases, respectively, are induced by the linear transformations with matrices

$$\begin{pmatrix} 1 & n \\ 0 & 1 \end{pmatrix} \quad \text{and} \quad \begin{pmatrix} 1 & 0 \\ -n & 1 \end{pmatrix}.$$

The derivatives of these Dehn twists are given by the classes of these matrices in $PSL(2, \mathbb{Z})$.

Remark. It is at this point that we use that there are only two curves. In fact, in this case, the α-atlas covers all of M and the homeomorphism is well-defined by its description in the α-atlas. Moreover, if there are more than two curves, the Dehn twist along α cannot in general be represented by an affine homeomorphism.

LEMMA 13.1. *An affine homeomorphism φ of M is a generalized pseudo-Anosov if and only if $D\varphi$ has real eigenvalues $(\lambda, 1/\lambda)$ with $\lambda \neq 1$.*

Proof. The condition on $D\varphi$ means that φ respects two transverse linear foliations by contracting the distances on the leaves of one and by stretching those of the other, by a constant factor. □

THEOREM 13.2. *Let $G(\alpha, \beta)$ be the subgroup of $A(M)$ generated by the affine Dehn twists along curves α and β that satisfy condition 13.1. The derivation map induces a homomorphism $D : G(\alpha, \beta) \to PSL(2, \mathbb{Z})$. Thus, $\varphi \in G(\alpha, \beta)$ is a generalized pseudo-Anosov if and only if $D\varphi$ has real eigenvalues distinct from ± 1. If, in addition, $\mathrm{card}(\alpha \cap \beta) = i(\alpha, \beta)$ (minimal intersection), then φ is pseudo-Anosov.*

Remark 1. In his announcement, Thurston says that there exists a homomorphism $G(\alpha, \beta) \to \mathrm{SL}(2, \mathbb{Z})$; this is a lift of D.

Remark 2. A matrix of $\mathrm{SL}(2, \mathbb{Z})$ whose trace has a modulus greater than 2 is Anosov. Thus, if φ is obtained by combining positive Dehn twists along α and negatives along β, with at least one of each, then φ is a generalized pseudo-Anosov.

Proof of Theorem 13.2. By Lemma 13.1, it only remains to prove the second assertion. Since a linear foliation has a spine if and only if the corresponding co-vertex is the center of a cell with two sides, the hypothesis of minimal intersection prohibits this configuration. □

Examples. In the first example, we take α to have a connected complement in M, a closed surface of genus 2. Let M' be the surface obtained by cutting M along α, and say that α_1 and α_2 are the two copies of α that form the boundary of M'. Four arcs joining α_1 to α_2 are needed to cut M' into two octagonal cells. But there is no way to glue α_1 to α_2 so that they make a connected curve. However, this becomes possible if each arc is doubled (see Figure 13.6). In this example, all the points of intersection are of the same sign, therefore the linear foliations are orientable; they have two singularities with four branches. They are thus defined by closed 1-forms with Morse saddles for singularities.

Other examples arise from the following lemma.

LEMMA 13.3. *Let α be a simple curve that is not homotopic to a point on the closed surface M. Then, there exists a simple curve β such that $M - (\alpha \cup \beta)$ is a union of cells. Moreover, if α is null-homologous, β can be chosen to be null-homologous.*

Proof. We find a decomposition of M into pairs of pants by curves K_j such that, for all j, we have $i(K_j, \alpha) \neq 0$ (if we think of α as a measured foliation, we can apply Lemma 6.16). Let β be the curve obtained by applying a (positive) Dehn twist along each K_j to α. We first prove that, for all $\gamma \in \mathcal{S}$, either $i(\alpha, \gamma) \neq 0$ or $i(\beta, \gamma) \neq 0$. Suppose that $i(\alpha, \gamma) = 0$; then, for some j, $i(\gamma, K_j) \neq 0$; otherwise γ would be isotopic to one of the K_ℓ and hence would intersect α. Now, applying Proposition A.1, we have

$$\left| i(\beta, \gamma) - \sum_j i(\alpha, K_j) i(\gamma, K_j) \right| \leq i(\alpha, \gamma) = 0.$$

Thus, $i(\beta, \gamma)$ is strictly positive.

This proves that any simple curve γ in $M - (\alpha \cup \beta)$ is nullhomotopic and therefore bounds a disk D in M. We see that D is contained in $M - (\alpha \cup \beta)$; otherwise, int D would contain a piece of β (or of α). As β does not intersect the boundary of D, we see that β would be entirely contained in D, which is absurd. From this, it is easy to see that the components of $M - (\alpha \cup \beta)$ are open disks. □

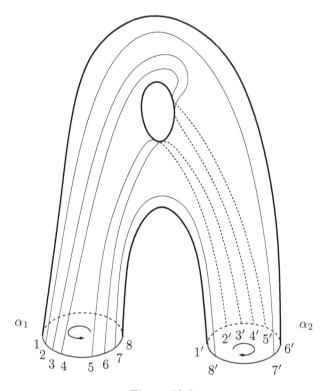

Figure 13.6.

When α and β are null-homologous, the affine homeomorphisms induce the identity on homology. However, some of them are not, up to isotopy, either periodic or reducible; this contradicts a conjecture of Nielsen [Nie44], saying that, if the eigenvalues of the induced automorphism on homology are on the unit circle, then the diffeomorphism is decomposable into periodic pieces.

Remark. All of the preceding constructions lead to pseudo-Anosovs where the dilatation factor is a quadratic integer. The members of the seminar do not know how to construct examples where it is of higher degree.

Exposé Fourteen

Fibrations over S^1 with Pseudo-Anosov Monodromy

by David Fried

We will develop Thurston's description of the collection of fibrations of a closed 3-manifold over S^1. We will then show that the suspended flows of pseudo-Anosov diffeomorphisms are canonical representatives of their nonsingular homotopy class, thus extending Thurston's theorem for surface homeomorphisms to a class of three-dimensional flows. Our proof uses Thurston's work on fibrations and surface homeomorphisms and our criterion for cross sections to flows with Markov partitions. We thank Dennis Sullivan for introducing Thurston's results to us. We are also grateful to Albert Fathi, François Laudenbach, and Michael Shub for their helpful suggestions.

A smooth fibration $f \colon X \to S^1$ of a manifold over the circle determines a nonsingular (i.e., never zero) closed 1-form $f^*(d\theta)$ with integral periods. Conversely, if ω is a nonsingular closed 1-form and X is closed, then the map $f(x) = \int_{x_0}^{x} \omega$ from X to $\mathbb{R}/\mathrm{periods}(\omega)$ will be a fibration over S^1 provided the periods of ω have rational ratios. For since $\pi_1(X)$ is finitely generated, the periods of ω will be a cyclic subgroup of $\mathbb{R}$ (not trivial since X is compact and f is open) and we have $\mathbb{R}/\mathrm{periods}(\omega) \cong S^1$. By constructing a smooth flow ψ on X with $\omega(d\psi/dt) = 1$, we see that f is a fibration. The relation of nonsingular closed 1-forms to fibrations over S^1 is very strong indeed, as the following theorem (which gives strong topological constraints on the existence of nonsingular closed 1-forms) indicates.

THEOREM 14.1 ([Tis70]). *For a compact manifold X, the collection $\mathcal{C}$ of nonsingular classes, that is, the cohomology classes of nonsingular closed 1-forms on X, is an open cone in $H^1(X; \mathbb{R}) - \{0\}$. The cone $\mathcal{C}$ is nonempty if and only if X fibers over S^1.*

Proof. The openness of $\mathcal{C}$ follows easily from de Rham's theorem. Indeed, if $\eta_1, \ldots, \eta_d$ are closed 1-forms that span $H^1(X; \mathbb{R})$ and if ω_0 is a closed 1-form, then the forms

$$\omega_a = \omega_0 + \sum_{i=1}^{d} a_i \eta_i, \quad |a_i| < \epsilon$$

represent a neighborhood of $[\omega_0]$ in $H^1(X; \mathbb{R})$. If ω_0 is nonsingular and ϵ is sufficiently small, then the ω_a are nonsingular. The forms $\lambda \omega_a$ with $\lambda > 0$ represent all positive multiples of $[\omega_a]$, so $\mathcal{C}$ is an open cone.

Choosing a so that the periods of ω_a are rationally related, we see that X fibers over S^1. We already noted that $0 \notin \mathcal{C}$. □

In dimension 3, Stallings characterized the elements of

$$\mathcal{C} \cap H^1(X; \mathbb{Z}) \subset H^1(X; \mathbb{R}).$$

We note that if X is closed, connected, and oriented and does fiber over S^1 with fibers of positive genus, then X will be covered by Euclidean space $\mathbb{R}^3$. Thus X will be *irreducible*, that is, every sphere S^2 embedded in X must bound a ball (this follows from Alexander's theorem that $\mathbb{R}^3$ is irreducible). We assume henceforward that M is a closed, connected, oriented, irreducible three-dimensional manifold.

THEOREM 14.2 ([Sta62]). *If $u \in H^1(M; \mathbb{Z}) - \{0\}$, then there is a fibration $f \colon M \to S^1$ with $[f^*(d\theta)] = u$, if and only if*

$$\ker(u : \pi_1(M) \to \mathbb{Z})$$

is finitely generated.

We observe that the forward implication holds even for finite complexes since the homotopy exact sequence identifies the kernel as the fundamental group of the fiber.

Theorem 14.2 reduces the geometric problem of fibering M to an algebraic problem, with only two practical complications. First, when $\dim H^1(M; \mathbb{R}) > 1$, there are infinitely many u to check. Second, it is difficult to decide if $\ker u$ is finitely generated. An infinite presentation may be readily constructed by the Reidemeister–Schreier process; this yields an effective procedure for deciding if the abelianization of $\ker u$ is finitely generated (we work out an example of this at the end of this exposé).

Thurston's theorem (Theorem 14.6 below) helps to minimize the first problem and make Stallings' criterion more practical. It will be seen that one need only examine finitely many u, provided one can compute a certain natural seminorm on $H^1(M; \mathbb{R})$.

As $H^1(M; \mathbb{Z}) \subset H^1(M; \mathbb{R})$ is a lattice of maximal rank, the seminorm will be determined by its values on $H^1(M; \mathbb{Z})$. Each $u \in H^1(M; \mathbb{Z})$ is geometrically represented by framed surfaces under the Pontrjagin construction [Mil65]. A framed (that is, normally oriented) surface S represents u whenever there is a smooth map $f \colon M \to S^1$ with regular value x so that $S = f^{-1}(x)$ and $u = [f^*(d\theta)]$. By irreducibility of M, any framed sphere in M represents the 0 class, so S may be taken to be *sphereless* (that is, all components of S have Euler characteristic less than or equal to 0).

14.1   THE THURSTON NORM

We set

$$\|u\| = \min\{-\chi(S) \mid S \text{ is a sphereless framed surface representing } u\}.$$

It is important to observe that a sphereless framed surface S in M with $\|u\| = -\chi(S)$ must be *incompressible* (that is, for each component $S_i \subset S$, the map $\pi_1(S_i) \to \pi_1(M)$ is injective). For (see Kneser's lemma [Sta71]) one could otherwise attach a 2-handle to S_i so as to lower $-\chi(S)$ without introducing spherical components.

The justification for the notation $\|u\|$ is the following result.

THEOREM 14.3 ([Thu86]). $\|u\|$ *is a seminorm on* $H^1(M;\mathbb{Z})$.

This follows from standard 3-manifold techniques. The triangle inequality follows from the incompressibility of minimal representatives and some cut and paste arguments. The homogeneity follows by the covering homotopy theorem for the cover $z^n \colon S^1 \to S^1$.

One instance where $\|u\|$ is easily computed is where u is represented by the fiber K of a fibration $f \colon M \to S^1$. We have:

PROPOSITION 14.4 ([Thu86]). *If* $K \to M \xrightarrow{f} S^1$ *is a fibration, then*

$$\|[f^*(d\theta)]\| = -\chi(K).$$

Proof. By homogeneity we may suppose that $u = [f^*(d\theta)]$ is indivisible, that is, $u(\pi_1(M)) = \pi_1(S^1)$. This implies that K is connected and that $K \times \mathbb{R}$ is the infinite cyclic cover of M determined by u. If K is a torus we are done, so assume $-\chi(K) > 0$. Any sphereless framed surface S representing u lifts to $K \times \mathbb{R}$, since for any component $S_0 \subset S$ we have $\pi_1(S_0) \subset \ker u = \pi_1(K)$. If $-\chi(S) = \|u\|$, then S is incompressible and $\pi_1(S_0) \to \pi_1(K \times \mathbb{R}) \cong \pi_1(K)$ is injective. Since subgroups of $\pi_1(K)$ of infinite index are free, we see that S_0 is a finite cover of K, hence

$$\|u\| = -\chi(S) \geq -\chi(S_0) \geq -\chi(K),$$

as desired.   □

In fact, we see that any sphereless framed surface S representing u with minimal $-\chi(S)$ is homotopic to the fiber K.

The behavior of $\|\ \|$ is decisively determined by the fact that integral classes have integral seminorms. We will show:

THEOREM 14.5 ([Thu86]). *A seminorm* $\|\ \| \colon \mathbb{Z}^n \to \mathbb{Z}$ *extends uniquely to a seminorm* $\|\ \| \colon \mathbb{R}^n \to [0,\infty)$. *A seminorm on* $\mathbb{R}^n$ *takes integer values on* $\mathbb{Z}^n$ *if and only if*

$$\|x\| = \max_{\ell \in F} |\ell(x)|,$$

where $F \subset \mathrm{Hom}(\mathbb{Z}^n, \mathbb{Z})$ *is finite.*

This enables us to state Thurston's description of the cone $\mathcal{C}$ of nonsingular classes, $\mathcal{C} \subset H^1(M;\mathbb{R}) - \{0\}$.

14.2   THE CONE $\mathcal{C}$ OF NONSINGULAR CLASSES

We will consistently use certain natural isomorphisms of the homology and co-homology groups of M. By the universal coefficient theorem,

$$H^1(M;\mathbb{Z}) \cong \mathrm{Hom}(H_1(M;\mathbb{Z});\mathbb{Z}) \quad \text{and}$$
$$H_1(M;\mathbb{Z})/\text{torsion} \cong \mathrm{Hom}(H^1(M;\mathbb{Z});\mathbb{Z}).$$

With real coefficients, $H^i(M;\mathbb{R})$ and $H_i(M;\mathbb{R})$ are dual vector spaces for any i. By Poincaré duality, we may identify $H^2(M;\mathbb{Z})$ with $H_1(M;\mathbb{Z})$. Thus we regard the Euler class χ_F of a plane bundle F on M, which is usually taken to be in $H^2(M;\mathbb{Z})$, as an element of $H_1(M;\mathbb{Z})$ and thus as a linear functional on $H^1(M;\mathbb{R})$.

THEOREM 14.6 ([Thu86]). $\mathcal{C}$ *is the union of (finitely many) convex open cones* $\mathrm{int}(T_i)$, *where* T_i *is a maximal region on which* $\| \ \|$ *is linear. The region* T_i *containing a given nonsingular 1-form* ω *is*

$$T_i = \{u \in H^1(M;\mathbb{R}) \mid \|u\| = -\chi_F(u)\}$$

where χ_F *is the Euler class of the plane bundle* $F = \ker \omega$.

Note. When $\| \ \|$ is a norm, we may say that $\mathcal{C}$ is all vectors $v \neq 0$ such that $v/\|v\|$ belongs to certain "nonsingular faces" of the polyhedral unit ball. Incidentally, we have that

$$\| \ \| \text{ is a norm} \quad \Longleftrightarrow \quad \text{all } T^2 \subset M \text{ separate } M$$
$$\Longleftrightarrow \quad \text{all incompressible } T^2 \subset M \text{ separate } M.$$

We now give our own analytic proof of Theorem 14.5.

Proof of Theorem 14.5. Clearly $\| \ \|$ extends by homogeneity to a seminorm $\| \ \|$ on $\mathbb{Q}^n$. This function is Lipschitz, hence has a unique continuous extension to a function $\mathbb{R}^n \to [0,\infty)$. The triangle inequality and homogeneity follow by continuity.

By convexity, all one-sided directional derivatives of $N(x) = \|x\|$ exist. Suppose $\tau = (0, \frac{1}{q}p)$ is a rational point, where $q \in \mathbb{Z}_+$ and $p = (p_2, \ldots, p_n) \in \mathbb{Z}^{n-1}$. For integral m, we compute

$$\begin{aligned}
\frac{\partial_+ N}{\partial x_1}(\tau) &= \lim_{m \to \infty} \frac{N(\tau + 1/qm\, e_1) - N(\tau)}{1/qm} \\
&= \lim_{m \to \infty} (N(1, mp_2, \ldots, mp_n) - N(0, mp_2, \ldots, mp_n)) \\
&\in \mathbb{Z},
\end{aligned}$$

since $\mathbb{Z}$ is closed.

By induction on n, we assume that $N(0, \bar{x})$, $\bar{x} \in \mathbb{R}^{n-1}$, is given by the supremum of finitely many functionals

$$\ell(\bar{x}) = a_2 x_2 + \cdots + a_n x_n, \quad a_2, \ldots, a_n \in \mathbb{Z},$$

$\bar{x} = (x_2, \ldots, x_n)$. By convexity, any supporting line L to

$$\mathrm{graph}(N) \subset \mathbb{R}^n \times \mathbb{R}$$

lies in a supporting hyperplane H (*supporting* means "intersects the graph without passing above it"). We choose $\bar{x}$ a rational point for which $N|0 \times \mathbb{R}^{n-1}$ is locally given by ℓ and choose L to pass through

$$(0, \bar{x}, N(0, \bar{x})) \in \mathbb{R}^n \times \mathbb{R}$$

in the direction of $(1, 0, \dfrac{\partial_+ N}{\partial x}(0, \bar{x}))$. Then we see that H is uniquely determined as the graph of

$$\left(\frac{\partial_+ N}{\partial x_1}(0, \bar{x}) \right) x_1 + a_2 x_2 + \cdots + a_n x_n.$$

So for a dense set of $\bar{x}$, the graph of N has a supporting functional at $(0, \bar{x})$ with integral coefficients.

Reasoning for each integrally defined hyperplane as we have for $\{x_1 = 0\}$, we find that integral supporting functionals

$$\ell(x) = a_1 x_1 + \cdots + a_n x_n, \quad a_i \in \mathbb{Z}$$

to the graph of N exist at a dense set in $\mathbb{R}^n$. Since N is Lipschitz, there is a bound $|a_i| \leq K, \quad i = 1, \ldots, n$. Thus the supporting functionals form a finite set F, so

$$S(x) = \sup_{\ell \in F} |\ell(x)|$$

is clearly a seminorm. But $S(x) \leq N(x)$ and equality holds on a dense set, implying that $S(x) = N(x)$ by continuity. $\square$

Before giving the proof of Theorem 14.6, let us observe one elementary consequence of Theorem 14.5. Since $\| \ \|$ is natural, any diffeomorphism $h \colon M \to M$ induces an isometry h^* of $H^1(M; \mathbb{R})$. If $\| \ \|$ is a norm, then the finite set of vertices of the unit ball spans $H^1(M; \mathbb{R})$ and is permuted by h^*.

COROLLARY 14.7. *If all incompressible $T^2 \subset M$ separate M, then the image of $\mathrm{Diff}(M)$ in $GL(H^1(M; \mathbb{R}))$ is finite.*

Proof of Theorem 14.6. Suppose ω and ω' are nonsingular closed 1-forms that are C^0–close. Then the oriented plane fields $F = \ker \omega$ and $F' = \ker \omega'$ are homotopic and so determine the same Euler class $\chi_{F'} = \chi_F \in H_1(M; \mathbb{R})$.

If $[\omega']$ is rational, let $q[\omega'] = \beta' \in H^1(M; \mathbb{Z})$, where $0 < q \in \mathbb{Q}$ and where β' is indivisible. Then if K' is the (connected) fiber of the fibration associated to $q\omega'$, we have

$$\chi(K') = \chi_{F'}(K') = \chi_F(K').$$

Using this and Proposition 14.4, we find

$$\|[\omega']\| = \frac{1}{q}(-\chi(K')) = -\frac{1}{q}\chi_F(K') = -\chi_F([\omega']).$$

Thus for all rational classes $[\omega']$ near $[\omega]$, the seminorm $\| \ \|$ is given by the linear functional $-\chi_F$. This shows that $\| \ \|$ agrees with $-\chi_F$ on a neighborhood of any nonsingular class $[\omega]$, as desired.

It only remains to show that every $\alpha \in \mathrm{int}(T)$ is a nonsingular class, where

$$T = \{\alpha \in H^1(M; \mathbb{R}) \mid \|\alpha\| = -\chi_F(\alpha)\}$$

is the largest region containing $[\omega]$ on which $\| \ \|$ is linear.

For this, we need a result of Thurston's thesis [Thu72] concerning the isotopy of an incompressible surface $S \subset M$ when M is foliated without "dead-end components." In fact, this result is explicitly stated for tori only, and one must see [Rou73] for a published account of this case. Restricting our attention to the foliation $\mathcal{F}$ defined by

$$\omega(\mathcal{F} \text{ is tangent to } \ker \omega = F),$$

we may state this result as follows: any incompressible, oriented, connected surface $S_0 \subset M$ with $-\chi(S_0) \geq 0$ may be isotoped so as to either lie in a leaf of $\mathcal{F}$ or to have only saddle tangencies with $\mathcal{F}$. (We call a tangency point s of S_0 with $\mathcal{F}$ a *saddle* if for some open ball B around s, the map

$$\int_s^x \omega \colon B \cap S_0 \to \mathbb{R}$$

has a nondegenerate critical point at s that is not a local extremum.)

Suppose $\alpha \in T \cap H^1(M; \mathbb{Z})$ is not a multiple of $[\omega]$. Represent α by a framed sphereless surface with $-\chi(S) = \|\alpha\|$. As S is incompressible, each component of S may be isotoped (independently) to a surface S_i that either lies in a leaf of $\mathcal{F}$ or has only saddle tangencies with $\mathcal{F}$. If some S_i lies in a leaf L of $\mathcal{F}$, then (as in Proposition 14.4) $\pi_1(S_i)$ would be of finite index in $\pi_1(L) = \ker[\omega]$. Since $\pi_1(S_i) \subset \ker \alpha$, we would find that α is a multiple of $[\omega]$. Thus each S_i has only saddle tangencies with $\mathcal{F}$.

LEMMA 14.8. *For each i, the normal orientations of S_i and $\mathcal{F}$ agree at all tangencies.*

Proof. We compute $\|\alpha\|$ in two ways. First,

$$\|\alpha\| = -\chi(S) = \sum_i -\chi(S_i).$$

Choosing some Riemannian metric on M, we may use the vector field V_i on S_i dual to $\omega|S_i$ to compute $-\chi(S_i)$. V_i will have only nondegenerate zeroes of index -1, since all tangencies are saddles. The Hopf index theorem [Mil65] gives $-\chi(S_i) = n_i$, where n_i is the number of tangencies of S_i with $\mathcal{F}$. Thus $\|\alpha\| = \sum n_i$.

On the other hand, we know that $\alpha \in T$ implies $\|\alpha\| = -\chi_F(\alpha)$. The natural normal orientations of F and S give us preferred orientations on F and

S_i, for each i. Each oriented plane bundle $F|S_i$ has an Euler class $\chi_F(S_i)[S_i]$ where $[S_i] \in H^2(S_i; \mathbb{Z})$ is the orientation class. We compute $\chi_F(S_i)$ as the self-intersection number of the zero section of $F|S_i$. For this purpose, look at the field W_i of vectors on S_i tangent to $\mathcal{F}$, which are the projections onto F of the unit normal vectors of S_i. Regarding W_i as a perturbation of the zero section of $F|S_i$, we compute the self-intersection number using the local orientations of F and S_i. When these orientations agree, one counts the singularity as -1 (just as in the tangent bundle case already considered) but when the orientations disagree one counts $+1$. Thus

$$-\chi_F(S_i) = n_i^+ - n_i^-,$$

where n_i^+ is the number of tangencies at which the orientations agree and n_i^- is the number of tangencies at which the orientations disagree. Thus

$$\|\alpha\| = \sum n_i^+ - \sum n_i^-.$$

Since $n_i = n_i^+ + n_i^-$, we have

$$\sum n_i^+ + \sum n_i^- = \|\alpha\| = \sum n_i^+ - \sum n_i^-,$$

whence all the nonnegative integers n_i^- must be zero. This proves the lemma.
□

Because of the lemma, we may define a framing N_i of S_i with $\omega(N_i) > 0$ everywhere. This framing may be extended to a product neighborhood structure on $U_i \supset S_i$, where

$$h \colon S_i \times [-1, 1] \to U_i$$

is a diffeomorphism, $h_*(\frac{\partial}{\partial t}) = N_i$ on $S_i = S_i \times 0$, and

$$\omega(h_*(\frac{\partial}{\partial t})) > 0.$$

Let $B \colon [-1, 1] \to [0, \infty]$ be a smooth function vanishing on $|x| > \frac{1}{2}$ with

$$\int_{-1}^{1} B = 1.$$

Letting $\eta_i = (\pi_2 h^{-1})^* B \, dt$ we find that, for all $s > 0$, we have

$$(\omega + s\eta_i)(h_*\frac{\partial}{\partial t}) > 0$$

on U. But since $\omega + s\eta_i = \omega$ away from U, we see that the closed 1-form $\omega + s\eta_i$ is nonsingular.

The portion of Theorem 14.6 already proven gives $[\omega + s\eta_i] \in \operatorname{int} T$. Thus,

$$[\eta_i] = \lim_{s \to \infty} \frac{[\omega + s\eta_i]}{s} \in T \cap H^1(M; \mathbb{Z})$$

for all i. So replacing $[\omega]$ by

$$[\omega] + s_1[\eta_1] + \cdots + s_{i-1}[\eta_{i-1}],$$

we see inductively that

$$[\omega] + s_1[\eta_1] + \cdots + s_i[\eta_i]$$

is nonsingular for all $s_1, \ldots, s_i \geq 0$. In particular, for all $s \geq 0$, we have that

$$[\omega] + s\alpha = [\omega] + s\sum[\eta_i]$$

is nonsingular.

We just showed that if $\beta = [\omega] \in \operatorname{int} T$ is a nonsingular class, then $\beta + s\alpha$ is nonsingular for all $\alpha \in T \cap H^1(M; \mathbb{Z})$ and $s \geq 0$. Now consider an arbitrary $\gamma \in \operatorname{int} T$, $\gamma \neq \beta$. By convexity we may find $v_1, \ldots, v_d \in \operatorname{int} T$, $d = \dim H^1(M; \mathbb{R})$, so that γ is in the interior of the d-simplex spanned by $\beta, v_1, \ldots, v_d$. We may choose $v_1, \ldots, v_d$ rational, say $v_j = \frac{1}{N}\alpha_j$, some $N \in \mathbb{Z}_+$, $\alpha_j \in \operatorname{int} T \cap H^1(M; \mathbb{Z})$. We have

$$\gamma = t_0\beta + \sum_{j=1}^{d} t_j\alpha_j,$$

with all $t_j > 0$. By induction on k, we see that each $\beta + \sum_{j=1}^{k}(t_j/t_0)\alpha_j$ is nonsingular. Setting $k = d$ and multiplying by $t_0 > 0$, we see that γ is nonsingular as well. Thus if one point $\beta \in \operatorname{int} T$ is nonsingular, all $\gamma \in \operatorname{int} T$ are nonsingular. $\square$

We will sharpen Thurston's theorem (Theorem 14.6) in the case when M is *atoroidal* (contains no incompressible embedded tori) and $H^1(M; \mathbb{Z}) \not\cong \mathbb{Z}$. We show (Theorem 14.11) that a nonsingular face T (i.e., one containing a nonsingular class) of the unit $\| \ \|$-ball determines a canonical flow $\varphi_t \colon M \to M$ such that $\operatorname{int} T$ consists precisely of all $[\omega]$ where ω is a closed 1-form with $\omega(\frac{\partial \varphi}{\partial t}) > 0$. We must begin by relating the atoroidal condition to Thurston's classification of surface homeomorphisms.

We suppose $f \colon M \to S^1$ is a fibration. Then flows ψ_t for which $\frac{d}{dt}f(\psi_t m) > 0$ for all m (we will only consider flows having a continuous time derivative) determine an isotopy class of surface homeomorphisms. For any $k \in K = f^{-1}(1)$, we consider the smallest time $T(k) > 0$ for which $\psi_{T(k)}(k) \in K$. This map $T(k) \colon K \to (0, \infty)$ is smooth (since the flow lines of ψ are transverse to K) and the *return map* $R(k) = \psi_{T(k)}(k)$ is a homeomorphism. By varying ψ, we obtain an isotopy class of homeomorphisms of the fiber K as return maps; this isotopy class will be called the *monodromy* of f, denoted $m(f)$.

We remark that the monodromy of f is determined algebraically by the cohomology class $\beta = f^*[d\theta] \in H^1(M; \mathbb{Z})$, or equivalently by the map $f_* \colon \pi_1(M) \to \pi_1(S^1)$. First assume that β is indivisible. From the exact homotopy sequence

$$1 \longrightarrow \pi_1(K) \longrightarrow \pi_1(M) \xrightarrow{f_*} \pi_1(S^1) \longrightarrow 1,$$

we see that $\pi_1(M)$ is the semidirect product $\pi_1(K) \ltimes_\alpha \mathbb{Z}$, where α is the outer automorphism of $\pi_1(K)$ determined by the monodromy of f. Thus $\pi_1(K)$ ($= \ker f_*$) and α are determined by f_* alone. Clearly the topological type of K is determined by $\pi_1(K)$; but Nielsen also showed that isotopy classes in $\mathrm{Diff}(K)$ correspond bijectively to outer automorphisms of $\pi_1(K)$. In general, $\beta = n\beta'$ is a positive integer multiple of an indivisible class β', and n is determined by $\mathrm{coker} f_* \cong \mathbb{Z}/n\mathbb{Z}$. We see that the fiber of f consists of n copies of K (where $\pi_1(K) = \ker f_*$) which are permuted cyclically by the monodromy. The nth power of the monodromy preserves K and acts on $\pi_1(K)$ by α (the outer automorphism of $\ker f_*$). Thus we may unambiguously speak of the monodromy of a nonsingular class $\beta \in H^1(M; \mathbb{Z})$.

We say that the monodromy $m(f)$ of a fibration $f \colon M \to S^1$ is *pseudo-Anosov* if the isotopy class has a pseudo-Anosov representative R. This representative is then uniquely determined within *strict conjugacy*, that is, for any two pseudo-Anosov representatives $R_0, R_1 \in m(f)$ there will be a homeomorphism g isotopic to the identity for which $R_0 g = g R_1$.

PROPOSITION 14.9. *Suppose that $H^1(M; \mathbb{Z}) \ncong \mathbb{Z}$. Given a fibration $f \colon M \to S^1$, M is atoroidal precisely when the monodromy $m(f)$ is pseudo-Anosov and the fibers of f are not composed of tori.*

Proof. Suppose M contains an incompressible torus S and let $\mathcal{F}$ be the foliation of M by the fibers of f. Again using the result of Thurston's thesis discussed in the proof of Theorem 14.6 [Rou73, Thu72], we may isotope S either to lie in a leaf of $\mathcal{F}$ or to be transverse to $\mathcal{F}$ (since $\chi(S) = 0$, the presence of saddle tangencies would force there to be tangencies of other types). If S does lie in a leaf, then the fibers of f are composed of tori parallel to S. If the torus S is transverse to $\mathcal{F}$, then one may define a flow ψ on M that preserves S and satisfies $\frac{d}{dt}(f \circ \psi_t) = 1$. Thus the return map $\psi_1 \colon K \to K$, $K = f^{-1}(1)$, preserves the family of curves $S \cap K$. Since S is incompressible, each of those curves is homotopically nontrivial in K. If the monodromy of f were pseudo-Anosov, these curves would grow exponentially in length under iteration by ψ_1. So we see that when $m(f)$ is pseudo-Anosov and the fibers of f are not unions of tori, then M must be atoroidal.

Conversely, when the fibers of f are unions of tori, these tori are essential. So we assume the components of the fibers have higher genus and that the monodromy is not pseudo-Anosov (hence reducible or periodic) and look for an incompressible torus. If $m(f)$ is reducible, we may construct ψ with $\frac{d}{dt}(f \circ \psi_t) = 1$ for which ψ_1 cyclically permutes a family of homotopically nontrivial closed curves $C \subset K$. Then $\{\psi_t C\}$ is an incompressible torus. If $m(f)$ has period n, after Nielsen (see Exposé 11), we may choose ψ with $\frac{d}{dt}(f \circ \psi_t) = 1$ for which $\psi_n = $ identity. Thus M is Seifert fibered. One may easily compute that $H^1(M; \mathbb{Z}) \cong \mathbb{Z}^{2g+1}$, where g is the genus of the topological surface that is the orbit space of ψ [Orl72]. As we assumed $H^1(M; \mathbb{Z}) \ncong \mathbb{Z}$, we must have a homologically nontrivial curve in this orbit space that corresponds to an incompressible torus in M. □

14.3 CROSS SECTIONS TO FLOWS

We may consider flows transverse to a fibration over S^1 from three viewpoints. The first is to begin with the fibration and produce transverse flows and an isotopy class of return maps. The second is to begin with a homeomorphism $R\colon K \to K$ and produce a fibration over S^1 with fiber K and a transverse flow φ with return map R. This is the well-known mapping torus construction, for which one sets

$$X = K \times [0,1]/(k,1) = (R(k),0),$$
$$f\colon X \to [0,1]/0 = 1 \cong S^1$$

the natural fibration and defines ψ to be the flow along the curves $k \times [0,1]$ with unit speed. Clearly $\psi_1|K \times 0 = R$ is the return map of ψ, as desired. This flow ψ is called the *suspension* of R. The third viewpoint is to begin with a flow ψ on X and to seek a fibration f over S^1 to which ψ is transverse—a fiber K is called a *cross section* to ψ. Note that K and ψ determine the return map R and an isotopy class of fibrations f.

In general, one has little hope of finding cross sections, since many manifolds do not fiber over S^1 at all. But there is a classification of the fibrations transverse to ψ which is especially concrete in the case of interest to us now.

Suppose that some cross section K to a flow φ has a return map $R\colon K \to K$ admitting a Markov partition $\mathcal{M} = \{S_1, \ldots, S_m\}$ (see Exposé 10 — the case we need is when R is pseudo-Anosov). There is a directed graph with vertices $S_1, \ldots, S_m$ and arrows $S_i \to S_j$ for each i and j for which $R(S_i)$ meets $\mathrm{int}(S_j)$. A *loop* ℓ for $\mathcal{M}$ is a cyclic sequence of arrows $S_{i_1} \to S_{i_2} \to \cdots \to S_{i_k} \to S_{i_1}$. Each loop ℓ determines a periodic orbit for R and thus a periodic orbit $\gamma(\ell)$ for φ. If all of $i_1, \ldots, i_k$ are distinct, we call ℓ *minimal*. There are only finitely many minimal loops ℓ.

We now discuss the classification and existence of cross sections to flows. Given a flow ψ on a compact manifold X there is a nonempty compact set of homology directions $D_\psi \subset H_1(X;\mathbb{R})/\mathbb{R}_+$, where the quotient space is topologized as the disjoint union of the origin and unit sphere. A *homology direction* for ψ is an accumulation point of the classes determined by long, nearly closed trajectories of ψ. We note that when K is a cross section to ψ, K is normally oriented by ψ and so determines a dual class $u \in H^1(X;\mathbb{Z})$. Let

$$\mathcal{C}_\mathbb{Z}(\psi) = \{u \in H^1(X;\mathbb{Z}) \mid u \text{ is dual to some cross section } K \text{ to } \psi\}.$$

THEOREM 14.10 ([Fri82b, Fri76]). *We have*

$$\mathcal{C}_\mathbb{Z} = \{u \mid u(D_\psi) > 0\}.$$

If φ, as above, has a cross section K and the return map R admits a Markov partition $\mathcal{M}$, then

$$\mathcal{C}_\mathbb{Z}(\varphi) = \{u \mid u(\gamma(\ell)) > 0 \text{ for all minimal loops } \ell \text{ for } \mathcal{M}\}.$$

Thus $\mathcal{C}_{\mathbb{Z}}(\psi)$ consists of all lattice points in a (possibly empty) open convex cone

$$\mathcal{C}_{\mathbb{R}}(\psi) = \{u \mid u(D_\psi) > 0\} \subset H^1(X; \mathbb{R}) - \{0\}.$$

It follows easily from Theorem 14.10 that

$$\mathcal{C}_{\mathbb{R}}(\psi) = \{[\omega] \mid \omega \text{ is a closed 1-form with } \omega(\tfrac{d\psi}{dt}) > 0\}.$$

Returning to our discussion of 3-manifolds, we call a flow φ on M *pseudo-Anosov* if it admits some cross section for which the return map is pseudo-Anosov. We now describe the cross sections to pseudo-Anosov flows, and show they are uniquely determined by their homotopy class among nonsingular flows on M.

THEOREM 14.11. *Suppose M fibers over S^1. Then each flow ψ on M that admits a cross section determines a nonsingular face $T(\psi)$ for the norm $\| \ \|$ on $H^1(M; \mathbb{R})$. Here*

$$T(\psi) = \{\|u\| = -\chi_{\psi^\perp}(u)\}$$

and $\psi^\perp$ denotes the normal plane bundle to the vector field $\frac{d\psi}{dt}$. One has $\mathcal{C}_{\mathbb{R}}(\psi) \subset$ int $T(\psi)$.

For any pseudo-Anosov flow φ on M, $\mathcal{C}_{\mathbb{R}}(\varphi) = $ int $T(\varphi)$.

The face $T(\varphi)$ (or the class $\chi_{\varphi^\perp}$) determines the pseudo-Anosov flow φ up to strict conjugacy. Thus any nonsingular face T on an atoroidal M with $H^1(M; \mathbb{Z}) \not\cong \mathbb{Z}$ determines a strict conjugacy class of pseudo-Anosov flows.

Proof. For $u \in \mathcal{C}_{\mathbb{Z}}(\psi)$, there is a cross section K to ψ dual to u. We have $\|u\| = -\chi(K)$, by Proposition 14.4. Since the restriction $\psi^\perp|K$ is the tangent bundle of K, we have $-\chi(K) = -\chi_{\psi^\perp}(u)$. Thus $-\chi_{\psi^\perp}$ is a linear functional on $H^1(M; \mathbb{R})$ that agrees with $\| \ \|$ on $\mathcal{C}_{\mathbb{Z}}(\psi)$ and the first paragraph of Theorem 14.11 is shown.

We now observe:

LEMMA 14.12. *Any cross section K to a pseudo-Anosov flow φ on M will have pseudo-Anosov return map R_K.*

Proof. By definition, there is some cross section L to φ with pseudo-Anosov return map R_L, but K and L will generally not be homeomorphic (one calls return maps to distinct cross sections to the same flow *flow-equivalent*). In any case, any structure on L invariant under R_L is carried over to a structure on K invariant under R_K under the system of local homeomorphisms between K and L determined by φ. This shows that R_K preserves a pair of transverse foliations $\mathcal{F}_K^u$ and $\mathcal{F}_K^s$ with the same local singularity structure as a pseudo-Anosov diffeomorphism.

We now show that the closure $\overline{P}$ of any prong P of $\mathcal{F}_K^u$ or $\mathcal{F}_K^s$ is the component K_0 of K that contains P. By passing to a cyclic cover $M_n \to M$ determined by the composite homeomorphism $\pi_1(M) \to (\pi_1(M)/\pi_1(K_0)) \cong \mathbb{Z} \to \mathbb{Z}/n\mathbb{Z}$ and restricting to the cross section $K_0 \subset M_n$ we may assume that K is connected and that R_K leaves P invariant (choose n so that P is invariant under $R_{K_0}^n$). Consider the closed R_L invariant subset $\{\varphi_t \overline{P}\} \cap L = I$. Since I contains the closure of a prong for the pseudo-Anosov diffeomorphism R_L, we know that I is dense in some component $L_0 \subset L$. As L_0 is a cross section to φ, we find that $\{\varphi_t \overline{P}\} = M$. As $\overline{P}$ is R_K invariant, we find $\overline{P} = K$ as desired.

Similarly we can check that the foliations $\mathcal{F}_K^u$ and $\mathcal{F}_K^s$ have no closed leaves.

It follows by the Poincaré–Bendixson theorem that each leaf closure contains a singularity, and thus a prong. So we find that all leaves of $\mathcal{F}_K^s$ and $\mathcal{F}_K^u$ are dense in their component of K.

We may see from this density of leaves and the fact that the local stretching and shrinking properties of R_K are the same as those of R_L that the Markov partition construction of Exposé 10 works for R_K. (It is easiest to construct birectangles for R_K by "analytic continuation" from immersed birectangles in L. This makes sense because K and L have the same universal cover.) As in the Anosov case [RS75], the Parry measures for the one-sided subshifts of finite type associated to $\mathcal{M}$ push forward to give transverse measures on $\mathcal{F}_K^u$ and $\mathcal{F}_K^s$ that transform under R_K by factors λ_K^{-1} and λ_K, for some $\lambda_K > 1$. As leaves are dense, these measures have positive values on any transverse interval but vanish on points. Thus R_K is pseudo-Anosov, and Lemma 14.12 is proved. □

Now suppose that φ^1 and φ^2 are pseudo-Anosov flows on M for which $\mathcal{C}_{\mathbb{R}}(\varphi^1)$ intersects $\mathcal{C}_{\mathbb{R}}(\varphi^2)$. Then we may choose

$$u \in \mathcal{C}_{\mathbb{R}}(\varphi^1) \cap \mathcal{C}_{\mathbb{R}}(\varphi^2) \cap H^1(M; \mathbb{Z})$$

and find fibrations $f_i \colon M \to S^1$ with $\dfrac{d}{dt}(f_i \circ \varphi_t^i) > 0$ and $u = [f_i^*(d\theta)]$, $i = 1, 2$.

As discussed earlier, u determines $m(f_i)$. This gives a homeomorphism $h \colon M \to M$ such that $f_1 \circ h = f_2$ where h acts on $\pi_1(M)$ by the identity. Thus h is isotopic to the identity [Wal68]. Hence, by this preliminary isotopy, we assume $f_1 = f_2 = f$ and denote the fiber by K.

Each φ^i determines a return map $R_i \colon K \to K$. By the lemma above, these R_i are pseudo-Anosov. Since the maps R_i are in the same isotopy class $h(f)$, they are strictly conjugate by the uniqueness of pseudo-Anosov diffeomorphisms (Exposé 12).

Now suppose that $gR_1 = R_2 g$, with g isotopic to the identity. Then the map $C_0 \colon M \to M$ defined by

$$C_0(\varphi_s^1 k) = (\varphi_s^2 gk), \quad k \in K, \ 0 \le s \le 1$$

is a homeomorphism conjugating flows φ^1 and φ^2 and $f \circ C_0 = f$. As $C_0|K = g$ is isotopic to the identity, C_0 may be isotoped to C_1 where $f \circ C_t = f$, for $t \in [0, 1]$ and C_1 fixes K. Since $\mathrm{Diff}(K)$ is simply connected [Ham66], we may isotope C_1 to the identity C_2 (through C_t satisfying $f \circ C_t = f$, $t \in [1, 2]$).

We have shown so far that if φ^i are pseudo-Anosov flows, $i = 1, 2$, then $\mathcal{C}_{\mathbb{Z}}(\varphi^1)$ either equals $\mathcal{C}_{\mathbb{Z}}(\varphi^2)$ or is disjoint from it, since conjugating a flow by conjugacy isotopic to the identity does not affect $\mathcal{C}_{\mathbb{Z}}$. It follows easily that the open cones $\mathcal{C}_{\mathbb{R}}(\varphi^1)$ and $\mathcal{C}_{\mathbb{R}}(\varphi^2)$ are either disjoint or equal.

Now suppose that φ is pseudo-Anosov but $\mathcal{C}_{\mathbb{R}}(\varphi)$ is a proper subcone of $\operatorname{int} T(\varphi)$. By Theorem 14.10, $\mathcal{C}_{\mathbb{R}}(\varphi)$ is defined by linear inequalities with integer coefficients, and so there is an integral class $u \in \operatorname{int} T \cap \partial \mathcal{C}_{\mathbb{R}}(\varphi)$. Then u is nonsingular (Theorem 14.6), the fibration corresponding to u has pseudo-Anosov monodromy (Proposition 14.9), and one obtains an Anosov flow ψ with $u \in \mathcal{C}_{\mathbb{R}}(\psi)$. This shows that $\mathcal{C}_{\mathbb{R}}(\psi)$ and $\mathcal{C}_{\mathbb{R}}(\varphi)$ are neither disjoint nor equal, contradicting the previous paragraph.

Thus we see that pseudo-Anosov flows satisfy $\mathcal{C}(\varphi) = \operatorname{int} T(\varphi)$. □

Theorem 14.11 shows that pseudo-Anosov maps satisfy an interesting extremal property within their isotopy class. Suppose $h_0 \colon K \to K$ has suspension flow $\psi_t^0 \colon M \to M$, where we take K connected and dual to the indivisible class $u \in H^1(M; \mathbb{Z})$. Given an isotopy h_t starting at h_0, we may deform ψ^0 through flows ψ^t with cross section K and return map h_t. We regard $u^{-1}(1)$ as a subset of $H_1(M; \mathbb{R})/\mathbb{R}_+$ and note that we always have $D_{\psi^t} \subset u^{-1}(1)$. By the Wang exact sequence

$$H_1(K; \mathbb{R}) \xrightarrow{(h_0)_* - \operatorname{Id}} H_1(K; \mathbb{R}) \longrightarrow H_1(M; \mathbb{R}) \xrightarrow{u} \mathbb{R} \longrightarrow 0,$$

we may identify $u^{-1}(1)$ with $u^{-1}(0) = \operatorname{coker}((h_0)_* - \operatorname{Id})$ by some fixed splitting of u. Whenever $h_s = h_t$, the simple connectivity of $\operatorname{Diff}(K)$ [Ham66] implies that $D_{\psi^s} = D_{\psi^t}$. Thus we may unambiguously associate a set of homology directions $D_h \subset \operatorname{coker}((h_0)_* - \operatorname{Id})$ to homeomorphisms h isotopic to h_0. Now assume that h_0 is pseudo-Anosov. By Theorem 14.11, we have

$$\mathcal{C}_{\mathbb{R}}(\psi^s) \subset \operatorname{int} T(\psi^s) = \operatorname{int} T(\psi^0) = \mathcal{C}_{\mathbb{R}}(\psi^0).$$

Thus we find, using Theorem 14.10, that the convex hull of D_{h_s} (which may be identified with the asymptotic cycles of ψ^s in this situation [Fri82a, Sch57]) always contains the convex polygon determined at $s = 0$. Thus we may say that pseudo-Anosov diffeomorphisms have the fewest generalized rotation numbers in their isotopy class.

We may analyze the topological entropy of the return maps R_K of the various cross sections K to a pseudo-Anosov flow $\mathcal{C}$. We parametrize these cross sections K by their dual classes $u \in H^1(M; \mathbb{Z})$ and define $h \colon \mathcal{C}_{\mathbb{Z}}(\varphi) \to (0, \infty)$ by $h([K]) = h(R_K)$, the topological entropy of R_K. We showed in [Fri82a] that $1/h$ extends uniquely to a homogeneous, downward convex function

$$1/h \colon \overline{\mathcal{C}_{\mathbb{R}}(\varphi)} \to [0, \infty]$$

that vanishes exactly on $\partial \mathcal{C}_{\mathbb{R}}(\varphi)$. Thus $h(u)$ may be defined for all $u \in H^1(M; \mathbb{R})$ in a natural way. The smallest value of h on $\operatorname{int} T \cap \{\|u\| = 1\}$ defines an interesting measure of the complexity of φ (or equivalently, by Theorem 14.11,

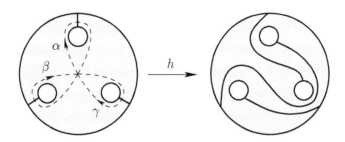

Figure 14.1.

of the face $T = T(\varphi)$). The integral points at which h is the largest give the "simplest" cross section to the flow φ (see [Fri82a]).

If one is given a pseudo-Anosov diffeomorphism $h\colon K \to K$ and a Markov partition $\mathcal{M}$ for h, Theorems 14.10 and 14.11 give an effective description of the nonsingular face T determined by the suspended flow $\varphi_t\colon M \to M$ of h, in terms of the orbits corresponding to minimal loops. As the computation of minimal loops in a large graph is difficult, we observe that there is a more algebraic way of using $\mathcal{M}$ to obtain a system of inequalities defining T. (We refer the reader to [Fri82a] for details, where we used this method to construct a rational zeta function for axiom A and pseudo-Anosov flows.) For sufficiently fine $\mathcal{M}$, we may associate to $\mathcal{M}$ a matrix A with entries in $H_1(M;\mathbb{Z})/\text{torsion} = H$. The expression $\det(I - A)$, regarded as an element in the group ring of the free abelian group H, may be uniquely written as $1 + \sum a_i g_i$, $g_i \in H - \{0\}$, $a_i \in \mathbb{Z} - \{0\}$, g_i distinct. Then T is defined by the inequalities $u(g_i) > 0$.

To illustrate Thurston's theory, it is convenient to work on a bounded M^3. The norm considered above can be extended to such M by omitting spheres and disks before computing the negative Euler characteristic. One should restrict to the case where ∂M is incompressible, and then Theorems 14.2 and 14.6 and Proposition 14.4 extend [Hem76, Thu86].

We let K be the quadruply connected planar region and h the indicated composite of the two elementary braids (Figure 14.1), which fixes the outer boundary component. We will let M be the mapping torus of h and compute $\| \ \|$. Rather than finding a pseudo-Anosov map isotopic to h, which would help compute only one face, we will instead compute $\ker(u\colon \pi_1(M) \to \mathbb{Z})$ for several indivisible $u \in H^1(M;\mathbb{Z})$. When this kernel is finitely generated, Theorem 14.2 shows u is nonsingular and Proposition 14.4 enables us to compute $\|u\|$. From a small collection of values of $\| \ \|$, Theorem 14.6 allows us to deduce all the others, indicating the existence of nonsingular classes that would be hard to detect using only Theorem 14.2.

We first compute $\pi_1(M) = \pi_1(K) \ltimes \mathbb{Z}$. Writing $\pi_1(K)$ as the free group on the loops α, β, and γ shown in the diagram, we find

$$
\begin{aligned}
\pi_1(M) \quad &= \quad \Big\langle \alpha, \beta, \gamma, t \,\Big|\, t^{-1}\alpha t = \gamma, \, t^{-1}\beta t = \gamma^{-1}\alpha\gamma, \\
&\qquad\qquad t^{-1}\gamma t = (\gamma^{-1}\alpha\gamma)\beta(\gamma^{-1}\alpha\gamma)^{-1} \Big\rangle \\
&= \quad \Big\langle \alpha, \beta, \gamma, t \,\Big|\, t^{-1}\alpha t = \gamma, \, t^{-1}\beta t = \gamma^{-1}\alpha\gamma, \, \gamma\beta t = \beta t \beta \Big\rangle \\
&= \quad \Big\langle \gamma, t \,\Big|\, (t\gamma^{-1}t\gamma t^{-1}\gamma)^2 = \gamma(t\gamma^{-1}t\gamma t^{-1}\gamma)t \Big\rangle.
\end{aligned}
$$

Abelianizing gives $H_1(M; \mathbb{Z}) \cong \mathbb{Z}\gamma \oplus \mathbb{Z}t$. Suppose $u \in H^1(M; \mathbb{Z})$ is indivisible, so that $a = u(\gamma)$ and $b = u(t)$ are relatively prime. The Reidemeister–Schreier process gives a presentation for $\ker(u \colon \pi_1(M) \to \mathbb{Z})$ (essentially by computing the fundamental group of the infinite cyclic covering corresponding to u) which is very ungainly for large a. When $a = 1$, one finds the relatively simple expression

$$
\ker u = \Big\langle t_i \,\Big|\, t_i\, t_{i+b-1}\, t_{i+b}^{-1}\, t_{i+b+1}\, t_{i+2b}\, t_{i+2b+1}^{-1} = t_{i+1}\, t_{i+b}\, t_{i+b+1}^{-1}\, t_{i+b+2} \Big\rangle.
$$

For $b > 1$, this relation expresses t_i in terms of $t_{i+1}, \ldots, t_{i+2b+1}$ and expresses t_{i+2b+1} in terms of $t_i, \ldots, t_{i+2b}$. Thus $\ker u$ is free on $t_1, \ldots, t_{2b+1}$. Similarly, if $b < -1$, then $\ker u$ is free on $t_1, \ldots, t_{1-2b}$ and if $b = 0$, then $\ker u$ is free on t_1, t_2, t_3. If $b = \pm 1$, however, one may abelianize and obtain

$$
(\ker u)^{\mathrm{ab}} = \mathbb{Z}[t, t^{-1}]/(2t^3 - 3t^2 + 3t - 2),
$$

which maps onto the collection of all 2^nth roots of unity, and so $\ker u$ is not finitely generated.

By Theorem 14.2, there is a fibration for $u = (1, b)$ when $b \neq \pm 1$, with fiber K_u satisfying $\pi_1(K_u) = \ker u$. By Proposition 14.4, $\|u\| = -\chi(K_u)$, which is clearly

$$
-1 + \mathrm{rank}(H_1(K_u)) = \begin{cases} |2b|, & b > 1, \, b \in \mathbb{Z} \\ 2, & b = 0. \end{cases}
$$

We will see that these values determine $\|\ \|$ completely. Using the dual basis to (γ, t), we know that

$$
\|(1, b)\| = \begin{cases} |2b|, & b > 1, \, b \in \mathbb{Z} \\ 2, & b = 0. \end{cases}
$$

But $\|(1, b)\|$ is a convex function f of b by Theorem 14.3 and it takes integer values at integer points. By convexity, $f(1)$ must be 2 or 3. Were $f(1) = 3$, convexity would force

$$
f(x) = \begin{cases} 2 + x & \text{for } 0 \leq x \leq 2 \\ 2x & \text{for } x \geq 2 \end{cases}
$$

and then $(1, 2)$ would not lie in an open face of the unit ball, contradicting Theorem 14.6. Thus one must have $f(1) = 2$, and likewise, $f(-1) = 2$. By convexity, we find $f(x) = \max(|2x|, 2)$. Homogenizing shows $\|(a, b)\| = \max(|2a|, |2b|)$, i.e., $\|u\| = \max(|u(2\gamma)|, |u(2t)|)$.

By Theorem 14.6, $u \in H^1(M; \mathbb{R})$ is nonsingular if and only if $|u(\gamma)| \neq |u(t)|$.

This example embeds in a larger one, constructed with the mapping torus M_0 of the transformation h^3 (M_0 is a triple cyclic cover of M). $H_1(M_0; \mathbb{Z})$ is free abelian on α, β, γ, t, so there is a norm on $H^1(M_0; \mathbb{R})$ whose restriction to

$$H^1(M; \mathbb{R}) \cong \{u \in H^1(M_0; \mathbb{R}) \mid u(\alpha) = u(\beta) = u(\gamma)\}$$

is $3\|\ \|$. We leave its computation as an exercise.

Exposé Fifteen

Presentation of the Mapping Class Group of a Compact

Orientable Surface

A Proof of a Theorem of Allen Hatcher and William Thurston

by François Laudenbach[1]; based on a lecture by Alexis Marin

15.1 PRELIMINARIES

Let M be a closed surface[2] of genus g and let G be the mapping class group of M (the elements of G are the isotopy classes of orientation-preserving diffeomorphisms of M). Let C be a multicurve in M with g components. We will say that C is a *marking* of M if $M - C$ is connected. The compact manifold with boundary obtained by cutting M along C is the disk Δ with $2g - 1$ holes. The group G acts transitively on the set of isotopy classes of markings via the right action

$$C \mapsto \varphi^{-1}(C),$$

where φ is a diffeomorphism of M. Let us choose a base marking C_0 and let us denote by H the subgroup of G stabilizing C_0. The group H is finitely presented: this is determined by decomposing H into three groups: the pure braid group on $2g - 1$ strands, the group of permutations of the components of C_0, and the group generated by the Dehn twists along each of the curves.

Hatcher and Thurston [HT80] have given a presentation of G modulo H. More precisely, they have constructed an element σ of G and words $\mu_1, \ldots, \mu_q$ whose letters belong to $\{\sigma^k : k \in \mathbb{Z}\} \cup H$ with the following properties:

1. H and σ generate G.

2. For $i = 1, \ldots, q$, the element m_i of G represented by μ_i belongs to H.

3. The words $\mu_i m_i^{-1}$ generate the relations of G; that is, conjugates of these elements generate the kernel of the natural homomorphism $H * \mathbb{Z} \to G$ associated to σ.

---

[1] I thank A. Marin for his oral exposition and for the clarifications that it brought to me on the work of Hatcher and Thurston.

[2] The case with boundary can be treated in an analogous fashion.

Even if one knows a presentation of H, this only says that there exists a presentation of G, but does not give it unless one knows how to calculate the m_i. It is true that the words μ_i are given by simple geometric constructions and that a diffeomorphism of Δ is entirely determined up to isotopy if one says what it does to some arcs. Thus, with enough courage, it is possible to make the "implicit relations" of Hatcher–Thurston into explicit ones.

Although the lecture by Marin reported faithfully on this work, it seems inappropriate to copy an article that has appeared. Instead, we will try to make the arguments of Hatcher–Thurston a little more conceptual. We will see, for example, in the proof of Lemma 15.4 a geometric fact particular to dimension 2 that contributes in an essential way to the finiteness. We have chosen not to give an explicit presentation of G, except in the case of the torus.

15.2 A METHOD FOR PRESENTING THE MAPPING CLASS GROUP

Let X be a simply-connected polyhedral complex of dimension 2 (possibly not locally finite), in which each edge and face is determined by its vertices. Let x_0 be a basepoint, let A be the set of edges containing x_0, and let F be the set of faces containing x_0. Suppose that the group G acts cellularly on X on the right, that G acts transitively on the 0-skeleton $X^{[0]}$, and that H is the stabilizer of x_0; then H acts on A and F on the right. We suppose that A/H and F/H are finite and we choose sets of representatives of each orbit, $a_1, \ldots, a_p$, and $f_1, \ldots f_q$.

Generators. We choose $\sigma_1, \ldots, \sigma_p \in G$ so that the two endpoints of a_i are $(x_0, x_0\sigma_i)$. Then the endpoints of any edge of X can be written as $(x_0 g, x_0\sigma_i hg)$, where $g \in G$, $h \in H$. A word $\sigma_{i_k} h_k \cdots \sigma_{i_1} h_1$ describes a path of edges starting from x_0 and passing successively through $x_1 = x_0\sigma_{i_1}h_1$, $x_2 = x_0\sigma_{i_2}h_2\sigma_{i_1}h_1$, etc. (see Figure 15.1).

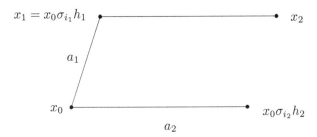

Figure 15.1. Points x_1 and x_2 are joined by $a_2 \cdot \sigma_{i_1}h_1$

Every path of edges has such a description. The connectedness of X implies that $\sigma_i, \ldots, \sigma_p$ generate G. Note that a word represents a loop if and only if the product of the letters belongs to H.

Relations. The action of G on X gives rise to three types of relations.

Backtracking. In expressing $(x_0, x_0\sigma_i, x_0)$ as a loop, we obtain a relation; that is, there exists an integer $j \in [1, p]$ and $h \in H$ such that

$$\sigma_j h \sigma_i \in H.$$

Different writings of the same edge. One must calculate the stabilizer T_i of the edge $(x_0, x_0\sigma_i)$ and, for each $t \in T_i$, write

$$(x_0, x_0\sigma_i)t = (x_0, x_0\sigma_i),$$

that is,

$$\sigma_i t \sigma_i^{-1} \in H.$$

Faces. The boundary of each face f_i, $i = 1, \ldots, q$, gives a relation.

To see that one thus obtains a presentation of G modulo H in the sense of Section 15.1, it suffices to recall that any homotopy of a loop of edges of X is formed from the following elementary operations: backtracking and insertion/deletion of the boundary of a face. We have therefore shown the following.

PROPOSITION 15.1. *Suppose that G acts on a cell complex X, and that it acts transitively on the vertices. Denote by H the stabilizer of some particular vertex x_0, and suppose that there are finitely many H-orbits of edges and faces passing through x_0 (in other words, the quotient of X by G is a finite complex with one vertex). If H is finitely presented and if the stabilizers of edges are of finite type,[3] then G is finitely presented.*

The objective now is to find a complex X on which the mapping class group acts. The first one that comes to mind is the nerve N of the space of C^∞ real-valued functions (of codimension at most 2) on M, given by its natural stratification[4] [Cer70]. The complex N is simply connected, but G does not act transitively on $N^{[0]}$. We can try considering only the codimension-1 strata that correspond to essential crossings (see below). We find then a simply-connected nerve where G acts transitively on the vertices, but the stabilizer of a vertex is bigger than H and seems difficult to study. We are nevertheless going to utilize these ideas to exhibit the set of isotopy classes of markings as the 0-skeleton of a complex whose simple connectivity follows from that of N. Lemma 15.2 is utilized for this purpose.

---

[3]We say that a group is of *finite type* if it has an Eilenberg–MacLane space with finitely many cells.

[4]The space of C^∞ functions with isolated critical points admits a natural stratification. The codimension 0 stratum consists of points where the critical values are all distinct, the codimension 1 stratum consists of points where exactly two critical values coincide, etc.

If Y and Z are two connected complexes, we will say that $\pi : Y \to Z$ is *cellular* if the following conditions are satisfied:

1. For each cell σ of Y, $\pi(\sigma)$ is a cell of Z.

2. The map $\pi|_{\text{int}\,\sigma}$ is a fibration onto its image (int σ denotes an open cell).

For example, if σ is a 2-cell, $\pi(\sigma)$ can be a point; or $\pi(\sigma)$ can be an edge, in which case $\partial\sigma$ is the union of four edges τ_1, τ_2, τ_3, τ_4, with $\pi(\tau_1) = 1$ point, $\pi(\tau_3) = 1$ point, and $\pi(\tau_2) = \pi(\sigma) = \pi(\tau_4)$; finally $\pi(\sigma)$ can be a 2-cell, in which case $\pi|_{\partial\sigma}$ is degree 1 onto its image. We very easily obtain the following.

LEMMA 15.2. *Let* $\pi : Y \to Z$ *be a cellular map in the above sense. We suppose that*

(i) for each $x \in Z^{[2]} - Z^{[1]}$, $\pi^{-1}(x)$ *is nonempty;*

(ii) for each $x \in Z^{[1]} - Z^{[0]}$, $\pi^{-1}(x)$ *is connected; and*

(iii) for each $x \in Z^{[0]}$, $\pi^{-1}(x)$ *is simply connected.*

Then $\pi_1(Z) = 1$ *implies* $\pi_1(Y) = 1$. *The converse is true as long as* $\pi^{-1}(x)$ *is connected for all* $x \in Z^{[0]}$ *and* π *is surjective on the 1-skeleton.*

15.3   THE CELL COMPLEX OF MARKED FUNCTIONS

We consider the space $\mathcal{F}$ of C^∞ functions on M, of codimension 0, 1, or 2, with the action by $\text{Diff}(M) \times \text{Diff}(\mathbb{R})$, and the nerve N of $\mathcal{F}$ stratified by the orbits. The mapping class group G acts on the right by the formula

$$f \mapsto f \circ \varphi,$$

where $f \in \mathcal{F}$, $\varphi \in \text{Diff}(M)$. Two elements of $\mathcal{F}$ are said to be *isotopic* if they are in the same orbit of the identity component of $\text{Diff}(M)$.

To any function $f \in \mathcal{F}$, we can associate its *graph of level sets* $\Gamma(f)$: the projection $M \to \Gamma(f)$ identifies two points if they belong to the same connected component of a level set of f; the quotient $\Gamma(f)$ is a complex of dimension 1. If f is generic, the Betti number $\beta_1(\Gamma(f))$ is equal to the genus g of M. If f is of codimension 1 and $\beta_1(\Gamma(f)) = g - 1$, we say that f belongs to a *stratum of essential crossings*. Figure 15.2 shows the critical level set at a crossing as well as the two neighboring level sets, and Figure 15.3 shows the graphs of the functions on a path crossing the stratum.

An edge of N that is dual to a stratum of an essential crossing is said to be of the *first type*. All other edges are of the *second type*. A *face* of N is said to be *principal* if it is dual to a stratum of equality of three critical values belonging to three strata of essential crossings; such a face is a hexagon that alternates three edges of the first type and three edges of the second type. Figures 15.4, 15.5, and 15.6, respectively, show the (immersed) level sets of a principal function of codimension 2, the corresponding stratification in the space of functions, and the graph of an unspecified neighboring generic function.

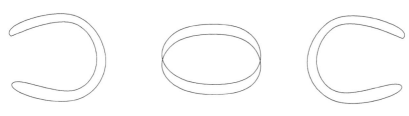

Figure 15.2. Level sets

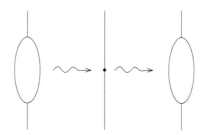

Figure 15.3.

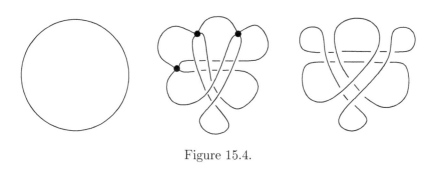

Figure 15.4.

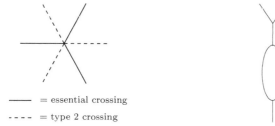

Figure 15.5.   Figure 15.6.

Marked functions. We say that (f, C) is a *marked function* if C is a marking of M where each component is contained in a level set of f. Each component of the marking corresponds to an edge of $\Gamma(f)$ and the complement of these (open) edges is a maximal subtree. All generic functions admit a marking, but a function of codimension 1 or 2 belonging to a stratum of essential crossings only admits an *incomplete marking* ($g - 1$ components).

For example, if f belongs to a stratum of an essential crossing, we can mark f by simple curves $\alpha_1, \ldots, \alpha_{g-1}$. If we mark the neighboring generic functions f' and f'', on both sides of the stratum, by $(\alpha_1, \ldots, \alpha_{g-1}, \alpha')$ and $(\alpha_1, \ldots, \alpha_{g-1}, \alpha'')$ respectively, then the minimal intersection of α' and α'' is one point.

We put the following isotopy relation on marked functions: (f, C) is isotopic to (f', C') if f is isotopic to f' and C is isotopic to C'. This relation is less fine than the relation of isotopy of pairs.

The 0-skeleton. The cell complex Y of marked functions is constructed with the set of isotopy classes of marked functions as the 0-skeleton. We have a projection

$$\pi : Y^{[0]} \to N^{[0]}$$

by forgetting the marking. The fiber above $[f] \in N^{[0]}$ is formed of all the markings of f up to isotopy.

LEMMA 15.3. *There exists a bound, independent of f, for the cardinality of* $Y^{[0]} \cap \pi^{-1}([f])$.

Proof. The graph $\Gamma(f)$ collapses onto an (uncollapsible) reduced subgraph $\Gamma_{\mathrm{red}}(f)$. The number of markings, up to isotopy of f, coincides with the number of markings of $\Gamma_{\mathrm{red}}(f)$. Indeed, if two markings C_1 and C_2 of f mark $\Gamma(f)$ on both sides of the foot of a collapsible tree (see Figure 15.7), then C_1 and C_2 are isotopic.

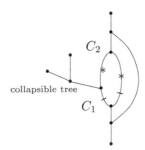

C_2

collapsible tree

C_1

Figure 15.7.

The lemma follows from the fact that, fixing the Betti number, there are only a finite number of reduced graphs up to piecewise-linear isomorphism. □

We remark that Lemma 15.3 is not true if we endow the marked functions with the finer isotopy relation.

The group G acts on $Y^{[0]}$ on the right by the formula

$$(f, C) \cdot [\phi] = (f \circ \varphi, \varphi^{-1}(C)).$$

This action is not transitive. The projection π is equivariant.

The 1-skeleton. We define three types of edges for the complex Y.

Edges of the first type. Let y_0 and y_1 be two vertices of Y represented by (f_0, C_0) and (f_1, C_1), where C_0 and C_1 have $(g - 1)$ components in common. To each edge of the first type of N joining $[f_0]$ to $[f_1]$ (unique if it exists),[5] we associate an edge of Y, said to be of the first type, between y_0 and y_1.

Observe that if such an edge exists, the incomplete marking common to C_0 and C_1 necessarily marks the function of codimension 1 from the essential crossing; the distinct components intersect in one point.

Edges of the second type. Let (f_0, C) and (f_1, C) be two vertices having the same marking. Let f_t, $t \in [0, 1]$, be a path representing an edge τ of the second type in N. If, up to isotopy, C is a marking of f_t for all t, we lift τ to an edge, said to be of the second type, from (f_0, C) to (f_1, C).

Edges of the third type. We join by an edge, said to be of the third type, each pair of distinct vertices of $\pi^{-1}([f])$.

The projection π and the action of G extend naturally to $Y^{[1]}$.

The 2-skeleton. The complex Y has three types of faces, as follows.

Faces of type I. By examining the geometric models associated to each stratum of codimension two of the space of functions ([Cer70]), we verify that, for each face σ of N, there exists a loop γ of $Y^{[1]}$ such that $\pi|_\gamma$ is an isomorphism of γ onto $\partial\bar{\sigma}$.

So in each G-equivalence class of the faces of N, we choose one representative $\bar{\sigma}$, we choose one lift γ of $\partial\bar{\sigma}$ satisfying the preceding condition, we attach to γ one 2-cell σ, and we saturate by the action of G.

Note that condition *(i)* of Lemma 15.2 is satisfied.

---

[5]If there were two edges of the first type from $[f_0]$ to $[f_1]$, we would have for f_1 two markings C_1 and C_1' such that $i(C_1, C_1') \neq 0$, which is absurd (here $i(\cdot, \cdot)$ is intersection in the sense of Thurston; see Exposé 4). To see this, we utilize the classification of crossings of critical values, due to J. Cerf [Cer70].

Faces of type II. Let τ and τ' be two edges of Y lifting the same edge of the first or second type of N. In joining their endpoints through fibers of π, we form a square or a triangle, onto which we attach a face σ. We extend π to σ, with values in $\pi(\tau) = \pi(\tau')$, in such a way that π is cellular.

Condition *(ii)* of Lemma 15.2 is now satisfied.

Faces of type III. To each triangle of the fiber of π, we attach a face by brute force; this makes the fibers $\pi^{-1}([f])$ simply connected for each $[f] \in N^{[0]}$, and so condition *(iii)* of Lemma 15.2 is satisfied.

Finally, we have constructed Y, which admits a cellular action by G, and which is simply connected by Lemma 15.2.

Remark. We could have skipped the edges of third type and, by consequence, the faces of types *II* and *III*. In this language, Hatcher and Thurston would have put an edge between (f, C_0) and (f, C_1) only if C_0 and C_1 have $n - 1$ components in common. The advantage of their restrained system is to obtain relations in G that are all carried by a surface of genus 2 with boundary.

15.4 THE MARKING COMPLEX

The 0-skeleton $X^{[0]}$ is formed from isotopy classes of markings of M. We have an equivariant projection
$$P : Y^{[0]} \to X^{[0]}$$
by forgetting the function. We recall that the group G acts transitively on $X^{[0]}$ with H as stabilizer.

Two distinct markings C_0 and C_1 are joined by an edge whenever there exist marked functions (f_0, C_0) and (f_1, C_1) joined by an edge of $Y^{[1]}$. The action of G extends to $X^{[1]}$ and the projection P extends equivariantly to $Y^{[1]}$. For example, if (f_0, C) and (f_1, C) are connected by an edge (necessarily of the second type), its projection is a point.

We attach a 2-cell σ to a loop γ of $X^{[1]}$ if there exists a loop $\tilde{\gamma}$ of $Y^{[1]}$ such that

(i) $\tilde{\gamma} = \partial\tilde{\sigma}$, $\tilde{\sigma} \in Y^{[2]}$, and

(ii) the restriction of P to $\tilde{\gamma}$ is degree 1 onto γ.

It is then easy to extend P to a homeomorphism $\mathrm{int}\,\tilde{\sigma} \to \mathrm{int}\,\sigma$. Further, G acts on X.

By examining the types of faces, we see right away that, for each face σ of Y, either $P(\partial\sigma)$ is an edge or a point, or $P(\partial\sigma)$ is a loop in $X^{[1]}$ and $P|\partial\sigma$ is of degree 1 onto its image (this is, for example, the case for the lifts in Y of the principal faces of N). It is then immediate that P extends to Y.

The projection π sends $P^{-1}([C])$ injectively to the nerve of a convex open set of the space of functions, namely the open set of functions that admit C for a marking. Thus $P^{-1}([C])$ is surely connected (similarly simply connected). Thus X is simply connected (Lemma 15.2).

We said at the start that G acts transitively on the set of markings. Let C_0 be a fixed base marking. The next goal is to show that there are finitely many H-orbits of edges and faces of X containing C_0.

LEMMA 15.4. *Let f be a function marked by C_0. The set of cells of Y passing through (f, C_0) projects to a finite subset of the set of cells of X modulo H.*

Proof. Let G_f be the stabilizer of $\pi(f, C_0) = [f]$. We will prove in fact a stronger lemma where we consider all the cells meeting $\pi^{-1}([f])$ and where we replace H by the subgroup $G_f \cap H$.

By Lemma 15.3, there is only a finite number, uniformly bounded, of cells of Y above a cell of N. Moreover, applied to the 0-cells, this says that $G_f \cap H$ is of finite index in G_f. The lemma is therefore reduced to the statement that, in N, only a finite number of cells pass through f modulo G_f. This fact does not correspond to a general property of the stratification of the space of functions on an arbitrary manifold. In dimension 2, it suffices to prove it for edges corresponding to crossings (double critical values) and for faces coming from triple critical values, because the cells passing through $[f]$ that are dual to the singularities "at the origin" are finite in number. The general fact is that a dual cell to a stratum of equality of two or three values is determined by a system of sheets[6] adapted to f (see [Cer70]). But for surfaces, the Dehn twists along curves of level sets of f represent elements of G_f and act on the system of sheets adapted to f, with finitely many orbits. □

We use the term *small loop* for a loop of $\Gamma(f)$ that is not null-homotopic and only passes through two branch points.

If m is a local maximum (resp. minimum) of f, we denote by $d(m)$ the minimal number of edges that one must traverse in order to descend (resp. climb) from the vertex of $\Gamma(f)$ corresponding to m to a small loop; if there are no small loops, $d(m)$ is not defined.

A function is said to be *minimal* if it has only one local maximum and one local minimum. A Morse function f, with distinct critical values, is said to be *almost minimal* in either of the following cases:

(a) The graph $\Gamma(f)$ does not have a small loop (in this case, f is minimal).

(b) The graph $\Gamma(f)$ has at least one small loop, and, for each non-absolute extreme m, we have $d(m) \leq 2$.

We remark that the graphs of the almost minimal functions form a finite set. The almost minimal functions are important because, in general, there are no principal faces of N passing through a given minimal function.

---

[6]In Cerf's original paper, which is in French, the term for sheet is "nappe."

LEMMA 15.5. *There exists only a finite number of isotopy classes of almost minimal functions marked by C_0.*

Proof. Starting from a minimal function, we can only give birth to a finite number of pairs of critical points if we want to stay in the space of almost minimal functions; up to isotopy, the number of possible choices for each birth is finite. As every almost minimal function is obtained by this process starting from a minimal function, it suffices to prove the lemma for minimal functions. In fact, we are going to prove that if, in addition to the marking, one is given the graph, endowed with its height function, and the position of the marking on this graph, then the function is determined up to isotopy.

For this, we must locate the figure 8 critical levels in the disk Δ obtained by cutting M along the marking. The marked graph indicates which holes of Δ are surrounded by each loop of the figure 8. Thus, starting from the lowest level set, the critical curves are placed one after the other, in a way that is unique up to isotopy. From this, the lemma is clear.   □

If f, marked by C_0, is not almost minimal, $\Gamma(f)$ contains an edge α with a free endpoint m such that $d(m)$ is either undefined or is greater than 2. Collapsing α amounts to the elimination of two critical points of f. Let f' be the endpoint of this path.

LEMMA 15.6. *For each cell σ of Y passing through (f, C_0), there exists a cell σ' passing through (f', C_0) such that $P(\sigma) = P(\sigma')$.*

Proof. Since the edge that is collapsed when we pass from f to f' is far from the small loops, the elimination is independent of all the essential crossings or changes of marking that one can do starting from (f, C_0). From this, the lemma is clear. Note that if σ moves around the saddle corresponding to the branching point of α, then $\dim(P(\sigma)) < \dim(\sigma)$ and we take σ' with $\dim \sigma' = \dim(P(\sigma))$. □

As an immediate corollary of the last three lemmas, we obtain that, modulo H, there are only finitely many cells of X passing through C_0.

THEOREM 15.7. *The mapping class group G of a surface is finitely presented.*

Proof. In order to apply Proposition 15.1, it remains to prove that the stabilizers of edges are of finite type. Let C_0 and C_1 be two markings of M that, within their isotopy classes, have the minimal number of points of intersection. Let H_0 and H_1 be the stabilizers of $[C_0]$ and $[C_1]$ in X. If these two vertices are joined by an edge, then $H_0 \cap H_1$ is the stabilizer of the edge. By Proposition 3.13, this group is identified with the connected components of the group of diffeomorphisms of M leaving C_0 and C_1 invariant. Up to permutations, it is related to the group of diffeomorphisms of a certain disk with holes, which is therefore of finite type. □

15.5   THE CASE OF THE TORUS

For the torus, we have the simplification that a function only admits a single marking up to isotopy. Therefore, we have

$$Y \cong N.$$

By Lemmas 15.4, 15.5, and 15.6, we obtain the classes of cells passing through $[C_0]$ in X in the following way. We consider a marked function (f_0, C_0) that is almost minimal, that is, where the graph $\Gamma(f_0)$ looks like Figure 15.8. Then we consider an essential crossing starting from f_0 (unique modulo G_{f_0}) that we determine by choosing a curve C_1 intersecting C_0 in one point. Next we consider a principal face passing through this edge; this is determined by choosing a curve C_2 such that $C_1 \cap C_2 = 1$ point and $C_0 \cap C_2 = 1$ point. In the particular case of the torus, we find that, fixing C_0 and C_1, there are exactly two possibilities for C_2 up to isotopy (Figure 15.9), denoted C_2' and C_2'', respectively. Thus, in the quotient of X by H, the vertex $[C_0]$ lies in one edge (C_0, C_1) and two triangles (C_0, C_1, C_2') and (C_0, C_1, C_2'').

Figure 15.8.

We denote by σ the $90°$ rotation of the torus about the point where C_0 and C_1 coincide; we have $C_1 = C_0\sigma$, and

$$\sigma^2 \in H. \tag{15.1}$$

Let ρ be the Dehn twist along C_0, so that $C_2' = C_1\rho$.

Then, we have $C_2'' = C_1\rho^{-1}$. Finally, the geometry of the torus implies that σ takes the edge (C_0, C_2'') to the edge (C_1, C_2'). Thus the path (C_0, C_1, C_2') is described by the word $\sigma\rho^{-1}\sigma$, whereas the edge (C_0, C_2') is described by $\sigma\rho$. This gives the relation

$$\sigma\rho^{-1}\sigma\rho^{-1}\sigma \in H. \tag{15.2}$$

We see immediately that the other cell gives

$$\sigma\rho\sigma\rho\sigma \in H. \tag{15.3}$$

But relation (15.3) follows from (15.2) and (15.1). Finally the stabilizer of an edge is trivial. Thus we have written a complete system of relations modulo H. To completely determine relation (15.2), we calculate the effect of the written element on C_1, and find that it takes C_1 to C_2'. Thus $(\sigma\rho^{-1})^3$ stabilizes the edge (C_0, C_1), and so

$$(\sigma\rho^{-1})^3 = 1.$$

Finally, H is generated by ρ and σ^2, with the commutation relation $[\sigma^2, \rho] = 1$.

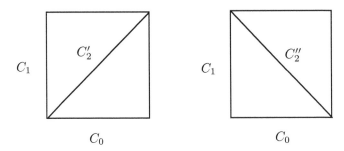

Figure 15.9.

Remark.   James McCool [McC75] has given a purely algebraic proof of the theorem. Joan Birman [Bir77], who was the first to give an explicit presentation in the case of a surface of genus 2, suggests that it seems difficult to exhibit a presentation from the proof of McCool. On the other hand, there is an approach using algebraic geometry (see [Mar77]).

Bibliography

[Abi80] William Abikoff, *The real analytic theory of Teichmüller space*, Lecture Notes in Mathematics, vol. 820, Springer, Berlin, 1980. (Cited on pages ix and xi.)

[AKM65] R. L. Adler, A. G. Konheim, and M. H. McAndrew, *Topological entropy*, Trans. Amer. Math. Soc. **114** (1965), 309–319. (Cited on pages 154 and 156.)

[Ano69] D. V. Anosov, *Geodesic flows on closed Riemann manifolds with negative curvature.*, Proceedings of the Steklov Institute of Mathematics, No. 90 (1967). Translated from the Russian by S. Feder, American Mathematical Society, Providence, RI, 1969. (Cited on page 7.)

[Bea83] Alan F. Beardon, *The geometry of discrete groups*, Graduate Texts in Mathematics, vol. 91, Springer-Verlag, New York, 1983. (Cited on page 114.)

[Ber78] Lipman Bers, *An extremal problem for quasiconformal mappings and a theorem by Thurston*, Acta Math. **141** (1978), no. 1-2, 73–98. (Cited on page xi.)

[Bes02] Mladen Bestvina, ℝ-*trees in topology, geometry, and group theory*, Handbook of geometric topology, North-Holland, Amsterdam, 2002, pp. 55–91. (Cited on page xi.)

[BH95] M. Bestvina and M. Handel, *Train-tracks for surface homeomorphisms*, Topology **34** (1995), no. 1, 109–140. (Cited on page xi.)

[Bir74] Joan S. Birman, *Braids, links, and mapping class groups*, Princeton University Press, Princeton, NJ, 1974, Annals of Mathematics Studies, No. 82. (Cited on page 15.)

[Bir77] _____, *The algebraic structure of surface mapping class groups*, Discrete groups and automorphic functions (Proc. Conf., Cambridge, 1975), Academic Press, London, 1977, pp. 163–198. (Cited on page 242.)

[BM77] Rufus Bowen and Brian Marcus, *Unique ergodicity for horocycle foliations*, Israel J. Math. **26** (1977), no. 1, 43–67. (Cited on page 191.)

[BO69]   R. L. Bishop and B. O'Neill, *Manifolds of negative curvature*, Trans.
          Amer. Math. Soc. **145** (1969), 1–49. (Cited on page 107.)

[Bon88]   Francis Bonahon, *The geometry of Teichmüller space via geodesic cur-
          rents*, Invent. Math. **92** (1988), no. 1, 139–162. (Cited on page xi.)

[Bou71]   J. P. Bourguignon, *Sur la structure du π_1, Séminaire de géométrie
          riemannienne 1970–71, variétés à courbure négative*, Séminaire de
          géometrie riemannienne 1970–71, variétés à courbure négative, 1971,
          Publications de l'Université Paris VII. (Cited on page 107.)

[Bow71]   Rufus Bowen, *Entropy for group endomorphisms and homogeneous
          spaces*, Trans. Amer. Math. Soc. **153** (1971), 401–414. (Cited on
          pages 156, 161 and 175.)

[Bow78]   ———, *Entropy and the fundamental group*, The structure of attrac-
          tors in dynamical systems (Proc. Conf., North Dakota State Univ.,
          Fargo, 1977), Lecture Notes in Mathematics, vol. 668, Springer,
          Berlin, 1978, pp. 21–29. (Cited on pages 161 and 162.)

[CB88]   Andrew J. Casson and Steven A. Bleiler, *Automorphisms of surfaces
          after Nielsen and Thurston*, London Mathematical Society Student
          Texts, vol. 9, Cambridge University Press, Cambridge, 1988. (Cited
          on pages x and xi.)

[CE75]   Jeff Cheeger and David G. Ebin, *Comparison theorems in Riemannian
          geometry*, North-Holland Mathematical Library, no. 9, North-Holland
          Publishing Co., Amsterdam, 1975. (Cited on pages 29 and 33.)

[Cer68]   Jean Cerf, *Sur les difféomorphismes de la sphère de dimension trois
          ($\Gamma_4 = 0$)*, Lecture Notes in Mathematics, vol. 53, Springer-Verlag,
          Berlin, 1968. (Cited on page 14.)

[Cer70]   ———, *La stratification naturelle des espaces de fonctions
          différentiables réelles et le théorème de la pseudo-isotopie*, Inst.
          Hautes Études Sci. Publ. Math. **39** (1970), 5–173.   (Cited on
          pages 233, 237 and 239.)

[Din71]   E. I. Dinaburg, *A connection between various entropy characteri-
          zations of dynamical systems*, Izv. Akad. Nauk SSSR Ser. Mat. **35**
          (1971), 324–366. (Cited on page 156.)

[DV76]   Adrien Douady and Jean-Louis Verdier, *Sur les formes de Strebel*,
          Séminaire de l'E.N.S., Astérisque, 1975–1976. (Cited on page 100.)

[EE69]   Clifford J. Earle and James Eells, *A fibre bundle description of Te-
          ichmüller theory*, J. Diff. Geom. **3** (1969), 19–43. (Cited on pages 107
          and 187.)

[Eps66]   D.B.A. Epstein, *Curves on 2-manifolds and isotopies*, Acta Math. **115** (1966), 83–107. (Cited on pages 2, 5 and 36.)

[Fen50]   W. Fenchel, *Remarks on finite groups of mapping classes*, Mat. Tidsskr. B. **1950** (1950), 90–95. (Cited on page 187.)

[FK65]   Robert Fricke and Felix Klein, *Vorlesungen über die Theorie der automorphen Funktionen. Band 1: Die gruppentheoretischen Grundlagen. Band II: Die funktionentheoretischen Ausführungen und die Andwendungen*, Bibliotheca Mathematica Teubneriana, Bände 3, vol. 4, Johnson Reprint Corp., New York, 1965. (Cited on page 107.)

[FLP79]   Albert Fathi, François Laudenbach, and Valentin Poénaru, *Travaux de Thurston sur les surfaces*, Astérisque, vol. 66, Société Mathématique de France, Paris, 1979, Séminaire Orsay, with an English summary. (Cited on page xi.)

[FM11]   Benson Farb and Dan Margalit, *A primer on mapping class groups*, Princeton University Press, Princeton, NJ, 2011. (Cited on pages xi and 77.)

[FN03]   Werner Fenchel and Jakob Nielsen, *Discontinuous groups of isometries in the hyperbolic plane*, de Gruyter Studies in Mathematics, vol. 29, Walter de Gruyter & Co., Berlin, 2003, Edited and with a preface by Asmus L. Schmidt, Biography of the authors by Bent Fuglede. (Cited on page 187.)

[Fri76]   David Fried, *Cross-sections to flows*, Ph.D. thesis, University of California, Berkeley, 1976. (Cited on page 224.)

[Fri82a]   ———, *Flow equivalence, hyperbolic systems and a new zeta function for flows*, Comment. Math. Helv. **57** (1982), no. 2, 237–259. (Cited on pages 227 and 228.)

[Fri82b]   ———, *The geometry of cross sections to flows*, Topology **21** (1982), no. 4, 353–371. (Cited on page 224.)

[Gan98]   F. R. Gantmacher, *The theory of matrices. Vol. 1*, AMS Chelsea Publishing, Providence, RI, 1998, Translated from the Russian by K. A. Hirsch, Reprint of the 1959 translation. (Cited on page 193.)

[Gau65]   Carl Friedrich Gauss, *General investigations of curved surfaces*, Translated from the Latin and German by Adam Hiltebeitel and James Morehead, Raven Press, Hewlett, NY, 1965. (Cited on page 1.)

[Goo71]   T.N.T. Goodman, *Relating topological entropy and measure entropy*, Bull. London Math. Soc. **3** (1971), 176–180. (Cited on page 175.)

[Gra73]   André Gramain, *Le type d'homotopie du groupe des difféomorphismes d'une surface compacte*, Ann. Sci. École Norm. Sup. (4) **6** (1973), 53–66. (Cited on page 186.)

[Gro00]   Mikhaïl Gromov, *Three remarks on geodesic dynamics and fundamental group*, Enseign. Math. (2) **46** (2000), no. 3-4, 391–402. (Cited on page 161.)

[Ham66]   Mary-Elizabeth Hamstrom, *Homotopy groups of the space of homeomorphisms on a 2-manifold*, Illinois J. Math. **10** (1966), 563–573. (Cited on pages 226 and 227.)

[Har77]   W. J. Harvey (ed.), *Discrete groups and automorphic functions*, Academic Press [Harcourt Brace Jovanovich Publishers], London, 1977. (Cited on page 179.)

[Hem76]   John Hempel, *3-Manifolds*, Princeton University Press, Princeton, NJ, 1976, Annals of Mathematics Studies, No. 86. (Cited on page 228.)

[HM79]   John Hubbard and Howard Masur, *Quadratic differentials and foliations*, Acta Math. **142** (1979), no. 3-4, 221–274. (Cited on page 100.)

[HT80]   A. Hatcher and William P. Thurston, *A presentation for the mapping class group of a closed orientable surface*, Topology **19** (1980), no. 3, 221–237. (Cited on page 231.)

[HT85]   Michael Handel and William P. Thurston, *New proofs of some results of Nielsen*, Adv. in Math. **56** (1985), no. 2, 173–191. (Cited on page xi.)

[Hub06]   John Hamal Hubbard, *Teichmüller theory and applications to geometry, topology, and dynamics. Vol. 1*, Matrix Editions, Ithaca, NY, 2006, Teichmüller theory, With contributions by Adrien Douady, William Dunbar, Roland Roeder, Sylvain Bonnot, David Brown, Allen Hatcher, Chris Hruska and Sudeb Mitra, With forewords by William Thurston and Clifford Earle. (Cited on page xi.)

[Iva92]   Nikolai V. Ivanov, *Subgroups of Teichmüller modular groups*, Translations of Mathematical Monographs, vol. 115, American Mathematical Society, Providence, RI, 1992, Translated from the Russian by E.J.F. Primrose and revised by the author. (Cited on page xi.)

[Kar66]   Samuel Karlin, *A first course in stochastic processes*, Academic Press, New York, 1966. (Cited on page 193.)

[Ker80]   Steven P. Kerckhoff, *The asymptotic geometry of Teichmüller space*, Topology **19** (1980), no. 1, 23–41. (Cited on page 100.)

[Lic64]   W.B.R. Lickorish, *A finite set of generators for the homeotopy group of a 2-manifold*, Proc. Cambridge Philos. Soc. **60** (1964), 769–778. (Cited on page 189.)

[Mal67]   B. Malgrange, *Ideals of differentiable functions*, Tata Institute of Fundamental Research Studies in Mathematics, No. 3, Oxford University Press, London, for the Tata Institute of Fundamental Research, Bombay, 1967. (Cited on page 34.)

[Man75]   Anthony Manning, *Topological entropy and the first homology group*, Dynamical systems—Warwick 1974 (Proc. Sympos. Appl. Topology and Dynamical Systems, Univ. Warwick, Coventry, 1973/1974; presented to E. C. Zeeman on his fiftieth birthday), Lecture Notes in Mathematics, vol. 468, Springer, Berlin, 1975, pp. 185–190. (Cited on page 161.)

[Mar77]   A. Marden, *Geometrically finite Kleinian groups and their deformation spaces*, Discrete groups and automorphic functions (Proc. Conf., Cambridge, 1975), Academic Press, London, 1977, pp. 259–293. (Cited on page 242.)

[McC75]   James McCool, *Some finitely presented subgroups of the automorphism group of a free group*, J. Algebra **35** (1975), 205–213. (Cited on page 242.)

[Mil65]   John W. Milnor, *Topology from the differentiable viewpoint*, Based on notes by David W. Weaver, University Press of Virginia, Charlottesville, 1965. (Cited on pages 216 and 220.)

[Mil68]   _____, *A note on curvature and fundamental group*, J. Diff. Geom. **2** (1968), 1–7. (Cited on pages 158 and 159.)

[Nie43]   Jakob Nielsen, *Abbildungsklassen endlicher Ordnung*, Acta Math. **75** (1943), 23–115. (Cited on page 186.)

[Nie44]   _____, *Surface transformation classes of algebraically finite type*, Danske Vid. Selsk. Math.-Phys. Medd. **21** (1944), no. 2, 89. (Cited on page 214.)

[Orl72]   Peter Orlik, *Seifert manifolds*, Lecture Notes in Mathematics, vol. 291, Springer-Verlag, Berlin, 1972. (Cited on page 223.)

[Orn74]   Donald S. Ornstein, *Ergodic theory, randomness, and dynamical systems*, Yale University Press, New Haven, CT, 1974, James K. Whittemore Lectures in Mathematics given at Yale University, Yale Mathematical Monographs, No. 5. (Cited on page 174.)

[Pal60] Richard S. Palais, *Local triviality of the restriction map for embeddings*, Comment. Math. Helv. **34** (1960), 305–312. (Cited on page 108.)

[Par64] William Parry, *Intrinsic Markov chains*, Trans. Amer. Math. Soc. **112** (1964), 55–66. (Cited on page 174.)

[Poé80] Valentin Poénaru, *Travaux de Thurston sur les difféomorphismes des surfaces et l'espace de Teichmüller*, Séminaire Bourbaki (1978/79), Lecture Notes in Mathematics, vol. 770, Springer, Berlin, 1980, Exp. No. 529, pp. 66–79. (Cited on page 1.)

[Rou73] Robert Roussarie, *Plongements dans les variétés feuilletées*, Publ. Math. I.H.E.S. **43** (1973), 143–168. (Cited on pages 220 and 223.)

[RS75] David Ruelle and Dennis Sullivan, *Currents, flows and diffeomorphisms*, Topology **14** (1975), no. 4, 319–327. (Cited on page 226.)

[Rus73] T. Benny Rushing, *Topological embeddings*, Pure and Applied Mathematics, vol. 52, Academic Press, New York, 1973. (Cited on page 127.)

[Sch57] Sol Schwartzman, *Asymptotic cycles*, Ann. Math. (2) **66** (1957), 270–284. (Cited on page 227.)

[See64] R. T. Seeley, *Extension of C^∞ functions defined in a half space*, Proc. Amer. Math. Soc. **15** (1964), 625–626. (Cited on page 34.)

[Sin76] Ya. G. Sinai, *Introduction to ergodic theory*, Mathematical Notes, no. 18, Princeton University Press, Princeton, NJ, 1976, Translated by V. Scheffer. (Cited on pages 8, 151 and 174.)

[Sma59] Stephen Smale, *Diffeomorphisms of the 2-sphere*, Proc. Amer. Math. Soc. **10** (1959), 621–626. (Cited on page 14.)

[Sma67] ———, *Differentiable dynamical systems*, Bull. Amer. Math. Soc. **73** (1967), 747–817. (Cited on page 7.)

[Smi34] P. A. Smith, *A theorem on fixed points for periodic transformations*, Ann. Math. (2) **35** (1934), no. 3, 572–578. (Cited on page 187.)

[Spr57] George Springer, *Introduction to Riemann surfaces*, Addison-Wesley Publishing Company, Reading, MA, 1957. (Cited on page 107.)

[Sta62] John Stallings, *On fibering certain 3-manifolds*, Topology of 3-manifolds and related topics (Proc. The Univ. of Georgia Institute, 1961), Prentice-Hall, Englewood Cliffs, NJ, 1962, pp. 95–100. (Cited on page 216.)

[Sta71]   ———, *Group theory and three-dimensional manifolds*, Yale Mathematical Monographs, vol. 4, Yale University Press, New Haven, CT, 1971, A James K. Whittemore Lecture in Mathematics given at Yale University, 1969. (Cited on page 217.)

[Ste69]   Shlomo Sternberg, *Celestial mechanics, part II*, Benjamin, New York, 1969. (Cited on page 10.)

[Thu72]   William P. Thurston, *Foliations of 3-manifolds that are circle bundles*, Ph.D. thesis, University of California, Berkeley, 1972. (Cited on pages 220 and 223.)

[Thu80]   ———, *The geometry and topology of 3-manifolds*, Unpublished manuscript, Princeton University Mathematics Department, 1980. (Cited on page 1.)

[Thu86]   ———, *A norm for the homology of 3-manifolds*, Mem. Amer. Math. Soc. **59** (1986), no. 339, i–vi, 99–130. (Cited on pages 217, 218 and 228.)

[Thu88]   ———, *On the geometry and dynamics of diffeomorphisms of surfaces*, Bull. Amer. Math. Soc. (N.S.) **19** (1988), no. 2, 417–431. (Cited on pages x, xi and 1.)

[Tis70]   D. Tischler, *On fibering certain foliated manifolds over S^1*, Topology **9** (1970), 153–154. (Cited on page 215.)

[Wal68]   Friedhelm Waldhausen, *On irreducible 3-manifolds which are sufficiently large*, Ann. Math. (2) **87** (1968), 56–88. (Cited on page 226.)

[Wey97]   Hermann Weyl, *Die Idee der Riemannschen Fläche*, Teubner-Archiv zur Mathematik. Supplement [Teubner Archive on Mathematics], vol. 5, B. G. Teubner Verlagsgesellschaft mbH, Stuttgart, 1997, Reprint of the 1913 German original, with essays by Reinhold Remmert, Michael Schneider, Stefan Hildebrandt, Klaus Hulek and Samuel Patterson, edited and with a preface and a biography of Weyl by Reinhold Remmert. (Cited on page 5.)

Index